▧▧▧▧▧ McGRAW-HILL BOOK COMPANY   NEW YORK   SAN FRANCISCO ▧▧▧▧▧
ST. LOUIS   TORONTO
LONDON   SYDNEY

ELEMENTS OF NUCLEAR PHYSICS

WALTER E. MEYERHOF

PROFESSOR OF PHYSICS

STANFORD UNIVERSITY

▨▨▨▨▨ TO MIRIAM AND ALEC ▨▨▨▨▨▨▨▨▨▨▨▨▨▨▨▨▨▨▨▨▨▨▨▨▨▨▨▨▨

PRINTED IN THE UNITED STATES OF AMERICA.

11 12 13 K P K P   7832109

ISBN 07-041745-8   LIBRARY OF CONGRESS CATALOG CARD NUMBER: 67-15247

⧼⧼⧼⧼⧼⧼⧼⧼⧼⧼⧼⧼⧼⧼⧼⧼⧼⧼⧼⧼⧼⧼⧼⧼⧼⧼⧼⧼⧼⧼ PREFACE ⧼⧼⧼⧼⧼

This book presents certain elements of nuclear physics at a level suitable for undergraduate physics students or for nuclear engineers. The material is also useful to scientists in other fields who wish to have more than a descriptive understanding of nuclear physics. The book grew out of a one-quarter course in nuclear physics for students who had as their physics preparation only a one-year college course, as well as a survey course in atomic physics, but whose mathematical preparation included calculus and ordinary differential equations.

The basic approach of the book is to present a limited amount of experimental information and to give the reader a feeling for its physical implications with the aid of quantum-mechanical concepts. Wherever possible, a preliminary discussion on the basis of classical theories is given. Sufficient quantum mechanics is introduced to permit correct order-of-magnitude estimates of nuclear quantities.

v

Beginning students of nuclear physics are often overwhelmed and discouraged by the wealth and diversity of the experimental and theoretical material. These tendencies have been avoided, first, by presenting only the important concepts in detail, and second, by giving as unified an approach as possible, based on the nuclear shell model. Also, comparisons are made between atomic and nuclear phenomena whenever they are helpful.

The book begins with a brief description of nuclear concepts. The next topic, nuclear structure, forms the heart of the subject. Those elements of quantum mechanics are given that are needed for an understanding of nuclear physics. Although radioactive decay and nuclear reactions reveal new aspects, they are treated, so far as possible, as extensions of the concepts of nuclear structure in order to emphasize the unity of the subject. The interactions of nuclear radiations with matter are briefly reviewed, because they are basic to the detection methods of nuclear radiations.

The material in the book is suitable for a one-semester course, and also for a one-quarter course if either the appendix on the two-nucleon system, the introduction of quantum mechanics, or the interaction of nuclear radiations with matter are omitted. The background given here should make it possible for the interested reader to pursue the study of nuclear accelerators, the applications of fission or fusion, and elementary-particle physics in more complete treatments. At the end of every chapter there are problems at various levels of difficulty that have been chosen to illustrate and elaborate the text. All the tabular or graphical material needed for their solution is included in the book.

The book has profited from several preliminary reviews, in particular by Professor W. E. Burcham, F.R.S., and from criticism by students. I owe the greatest indebtedness and gratitude to Marion Middleton for her skillful and accurate preparation of the manuscript and prior preliminary editions, and to my wife for her unbounded patience.

WALTER E. MEYERHOF

CONTENTS

BASIC NUCLEAR CONCEPTS

1-1 INTRODUCTION

A study of nuclear physics centers around two main problems. First, one hopes to understand the properties of the force which holds the nucleus together. Second, one attempts to describe the behavior of systems of many particles, such as nuclei are. These problems are related, since the properties of a system of many particles are to a large extent determined by the force that binds the particles together. But other aspects of such a system come about simply because many particles are interacting.

Physicists can discuss many-particle systems only within certain approximations, which are determined by the particular experimental fact they wish to explain. For example, it is often sufficient to discuss the behavior of a certain amount of gas in terms of the gas laws (Boyle's law, Charles' law); but this omits details of molecular motion which one needs to describe in order to understand the heat conductivity of a gas. In the case of nuclei, the approximate descriptions are called *models*. Much of the discussion in this book is based on such models, each one suited only for a limited range of experimental situations.

Although the historical development of nuclear physics will not be followed, a few of the highlights are presented in Table 1-1.

TABLE 1-1   Some of the highlights in the development of nuclear physics

Discovery of radioactivity (Becquerel)	1896
Rutherford's atomic model	1911
Discovery of isotopes (J. J. Thomson)	1912
Induced nuclear transmutation (Rutherford)	1919
Application of quantum mechanics to radioactivity:	
Alpha decay (Gamow, Gurney, and Condon)	1928
Beta decay (Fermi)	1934
Discovery of neutron (Chadwick)	1932
n-p hypothesis (Heisenberg)	1932
Discovery of positron (Anderson)	1932
Role of mesons in nuclear forces (Yukawa)	1935
Discovery of μ meson (Anderson and Neddermeyer)	1936
Discovery of π meson (Powell)	1946
Nonconservation of parity in beta decay (Lee and Yang)	1956

Becquerel[1] (1896) is generally credited with the discovery of radioactivity. This occurred when he noticed the accidental blackening of a photographic plate adjacent to a certain mineral. Pierre and Marie Curie (1898) succeeded in chemically separating the radioactive material (radium) from the ore. The greatest understanding of radioactivity was achieved by Rutherford and collaborators. They proposed that radioactivity should produce a change in the chemical species (1903) and investigated in detail the nature of the radiations. Three types of radiation were discovered, called alpha, beta and gamma. Once it was shown that alpha radiation consists of ionized helium atoms, the stage was set for Rutherford's interpretation of the alpha-particle scattering experiments of Geiger and Marsden (1909). Rutherford (1911) demonstrated that the

[1] References to original papers can be found in the bibliography at the end of this book.

scattering experiments could be explained only by assuming an atom consists of a massive, positively charged nucleus, of diameter ($\approx 10^{-12}$ cm) much smaller than the atomic diameter ($\approx 10^{-8}$ cm), surrounded by electrons. (In a neutral atom, the number of electrons is equal to the number of positive charges carried by the nucleus.) The first consistent model of the motion of the atomic electrons was accomplished by Bohr (1913).

Details of the nuclear constitution became clearer once the neutron had been discovered by Chadwick (1932), leading to Heisenberg's hypothesis (1932) that nuclei consist of protons and neutrons. At that time, too, attempts were made to understand the nuclear force. Experimentally, the force was found to be much stronger than any force then known, such as the electrical or gravitational force, and it also had a much shorter range. Taking up a suggestion by Heisenberg that the nuclear force is caused by an exchange of particles between nuclear constituents, Yukawa (1935) showed that if the exchanged particles are heavy enough the main features of the force could be explained. These particles, now called mesons, were later discovered in cosmic radiation.[1]

At present the main problems of nuclear physics, mentioned at the beginning of this section, are solved in broad outline, although not in detail. We know what properties the nuclear force possesses—it turns out to be a very complicated force. We also have learned how to relate the important features of nuclear models to the force. Yet many theoretical problems remain open. Experimentally, unexpected aspects of nuclei are discovered as the tools of research become more refined.

1-2 BASIC NUCLEAR PROPERTIES

Nuclei have certain time-independent properties such as mass, size, charge, intrinsic angular momentum (often called *nuclear spin*), and certain time-dependent properties such as radioactive decay and artificial transmutations (nuclear reactions). The nuclei also have excited states, whose energy is usually treated under the first class of properties, but whose decay is one of the types of radioactive decay. For an overall view of the field, each of the properties will be examined briefly. In later chapters more details will be given.

1-2a Nuclear mass and charge. Early chemical methods of mass comparison had already brought out the following approximate relation (Prout, 1815):

$$M \approx \text{integer} \times M_\text{H} \qquad (1\text{-}1)$$

where M = mass of a specific atom
M_H = mass of a hydrogen atom
The integer is now called *mass number* and will be denoted by the symbol A. It was shown by x-ray scattering (Barkla, 1911) that the number Z of atomic electrons, and hence the number of positive nuclear charges, was not equal to

---

[1] A more extensive historical account of the development of nuclear physics can be found in Burcham, 1963, sec. 1-1.

the mass number A. This made plausible the first hypothesis of nuclear structure, that nuclei consist of A protons and $A - Z$ bound electrons. As mentioned above, though, the discovery of the neutron (Chadwick, 1932) led Heisenberg (1932) to suggest that protons and neutrons are the fundamental constituents of all nuclei. The evidence for this is now beyond doubt, but can be understood only on the basis of quantum mechanics. One decisive example will be mentioned below. With the neutron-proton hypothesis we expect the mass of an atom to be

$$M \approx ZM_{\mathrm{H}} + NM_n \qquad (1\text{-}2)$$

where $Z =$ number of protons in nucleus (*atomic number*)
   $N =$ number of neutrons in nucleus (*neutron number*)
   $M_n =$ mass of a neutron

The discovery by Thomson (1912) of atomic species with identical chemical properties but different masses (called isotopes) stimulated the development of precise determinations of atomic or nuclear masses. This specialized branch of nuclear physics, pioneered by Aston (1919), is known as mass spectrometry. Its importance lies in the fact that a considerable amount of information about nuclear forces and nuclear structure can be obtained from precise mass measurements. This will be discussed in Chap. 2. We will see that there is a difference between the left and right sides of Eq. (1-2), which represents the nuclear binding energy.

1-2b Nuclear size.   The first detailed model of an atom, going beyond the kinetic theory (solid sphere) model, was proposed by J. J. Thomson (ca. 1900) soon after his discovery of atomic electrons. The electrons were assumed to float among massive positive charges of atomic dimensions ($\approx 10^{-8}$ cm). According to this model any high-speed particle could penetrate solid matter only by a diffusion process. On the other hand, scattering experiments of alpha particles by gold foils (Geiger and Marsden, 1909) showed a much larger amount of back scattering than a diffusion process would allow. Rutherford noticed that this implied the existence of a very small ($\ll 10^{-8}$ cm) atomic nucleus, exerting a simple electrical (coulomb) force on the alpha particle. He deduced the law of scattering.[1] Later measurements showed that this law is not obeyed if:

   1   The alpha-particle kinetic energy is too high.
   2   The atomic number of the scatterer is too low.

The critical energy T_α and corresponding atomic number Z, at which the scattering law breaks down, allow a rough estimate of the nuclear radius of the scatterer. We have to assume, if the distance of separation between the alpha particle and the center of the scatterer becomes smaller than this radius, nuclear forces come into play which are much stronger than the coulomb force used to derive the scattering law.

[1] A brief derivation is given in Sec. 5-4c.

When an alpha particle is very distant from a given nucleus, it has only kinetic energy T_α. It comes closest to the nucleus in a head-on collision. At that point, the alpha particle has only potential energy if the recoil of the nucleus is neglected. Hence, by conservation of energy,

$$T_\alpha = \frac{2eZe}{D} \qquad \text{(in electrostatic units)} \qquad (1\text{-}3)$$

where $2e$ = charge of the alpha particle ($e = 4.80 \times 10^{-10}$ esu)[1]
$\quad\, Ze$ = charge of the scattering nucleus
$\quad\; D$ = distance of closest approach

$$D = \frac{2Ze^2}{T_\alpha} \qquad\qquad (1\text{-}4)$$

For example, alpha particles show deviations from pure coulomb scattering on uranium beyond 25 Mev (1 Mev $= 1.60 \times 10^{-6}$ ergs).[2] In that case

$$D = \frac{2 \times 92 \times (4.8 \times 10^{-10})^2}{25 \times 1.6 \times 10^{-6}}$$

$$\approx 10^{-12} \text{ cm} = 10 \text{ F} \qquad (1 \text{ F} = 1 \text{ fermi} = 10^{-13} \text{ cm})$$

More refined experiments, using the scattering of other nuclear particles and of electrons, have shown that the radius at which nuclear effects occur can be written approximately

$$R = R_0 A^{1/3} \qquad\qquad (1\text{-}5)$$

where R_0 is called the *radius constant* and has the values

$$R_0 \approx \begin{cases} 1.4 \text{ F} & \text{for nuclear particle scattering on nuclei} \\ 1.2 \text{ F} & \text{for electron scattering on nuclei} \end{cases} \qquad (1\text{-}6)$$

The difference between these two values comes about as follows: In electron scattering we determine the location of the positive (point) charges associated with the protons in the nucleus. In nuclear-particle scattering we determine the size of the nuclear-force-producing region affecting the particle. It turns out that the nuclear force extends beyond the region with which charge (or mass) are associated, making the nucleus appear larger than it actually is. The force extension beyond nuclear matter is about 1 F and is determined by the range of the nuclear force.

The simple form of Eq. (1-5) would be obtained if the nucleus were a spherical assembly of A hard particles. In that case, the volume of the nucleus would be proportional to A and the radius proportional to $A^{1/3}$. This simple model, though correct in some respects, is oversimplified. Refined electron scattering experiments (Hofstadter et al., 1953) show that the nuclear density distribution

---

[1] For accurate values of certain physical constants, see Appendix D.
[2] Accurate values of certain conversion factors are listed in Appendix D.

does not have a sharp cutoff at the radius R, but has roughly the shape given in Fig. 1-1. Nevertheless the concept of *nuclear radius* is often useful. Equation (1-5) applied to U^{238} gives $R = 9\,F$ which compares favorably with the estimate provided by D from expression (1-4).

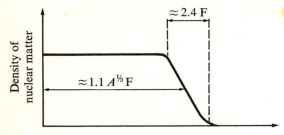

FIGURE 1-1 Density distribution of nuclear matter in a nucleus.

1-2c Intrinsic angular momentum of a nucleus.   The *angular momentum*[1] of a nucleus is an important quantity because, as we will see, it restricts the structure of complex nuclei and affects all dynamical nuclear properties. Only a few details of the angular momentum of a system of particles will be discussed in this section.

It is found experimentally and incorporated in the laws of quantum mechanics that neutrons and protons have an intrinsic angular momentum $\frac{1}{2}\hbar$, like electrons. ($\hbar$ is Planck's constant h divided by 2π.) Since angular momentum is a vector, the total angular momentum of a nucleus is the vector sum of the angular momenta of its constituents. We find, *experimentally*, that complex nuclei have angular momenta equal to $I\hbar$, where

For even-A nuclei: I is an integer (including zero)

For odd-A nuclei: I is an integer (including zero) plus one-half

For example, the nucleus of deuterium H^2 has $I = 1$ and the nucleus of Li^7 has $I = \frac{3}{2}$.

According to the quantum mechanical laws of addition of angular momenta, any system of P particles can have an angular momentum (about its center of mass) equal to an integer $\times\,\hbar$ if P is even, and an integer plus one-half $\times\,\hbar$ if P is odd. This applies to atomic electrons as well as to nuclear constituents. Therefore, if the nucleus H^2 were made up of two protons plus one electron (to give $Z = 1$), we would expect $I = \frac{1}{2}$ or $\frac{3}{2}$. If, on the other hand, it consists of one proton and one neutron, we expect $I = 0$ or 1. The latter value is in accord with experiment. The same reasoning extended to other nuclei shows that

---

[1] Angular momentum is defined in footnote 2 at the end of Sec. 2-2a. The nuclear angular momentum is often called *nuclear spin*, even though it has orbital, as well as intrinsic spin, contributions.

nuclei cannot consist of protons and electrons but must consist of protons and neutrons.[1]

We have not indicated how I is measured. Both atomic and molecular spectra are slightly influenced by magnetic effects due to the nuclear angular momentum, and the value of I can often be inferred.[2] Nuclear transmutations also are strongly affected by the angular momenta of the initial and final systems because they have to satisfy the law of conservation of angular momentum. This allows a determination of I in certain cases.

1-2d Dynamic properties of nuclei.   Nuclei, like atoms, can be in *excited states* of definite energies. Transitions between excited states occur by emission of electromagnetic radiation (gamma rays) completely analogous to light emission from atoms. The main difference is that, whereas atomic states are separated by energies of the order of an electron volt, the separations between nuclear states are about 10^4 to 10^6 ev. Just as a study of atomic spectra allows a reconstruction of atomic energy levels, which in turn has led to atomic models, a study of gamma-ray spectra leads to nuclear energy states and nuclear models.

Nuclei can also be *transformed* into each other. Some of the transformations occur spontaneously by the emission of positive or negative electrons (beta rays) or alpha particles. Other transformations can be induced by nuclear bombardments. In all cases the total number of nucleons is conserved. Furthermore, there are overall conservation of mass and energy, conservation of linear momentum, and conservation of angular momentum. No contradictions to these conservation laws have been found. They play an important role in most aspects of nuclear physics.

1-2e Nomenclature.   As in any specialized field, a certain nomenclature has developed based on convenience and tradition. The important terms are given below.

Nuclide   A specific nuclear species, with a given proton number Z and neutron number N

Isotopes   Nuclides of same Z and different N

Isotones   Nuclides of same N and different Z

Isobars   Nuclides of same mass number A $(A = Z + N)$

Isomer   Nuclide in an excited state with a measurable half-life

Nucleon   Neutron or proton

Mesons   Particles of mass between the electron mass (m_0) and the proton mass (M_H). The best-known mesons are π mesons $(\approx 270m_0)$, which play an important role in nuclear forces, and μ mesons $(207m_0)$ which are important in cosmic-ray phenomena.

---

[1] For a summary of other arguments in favor of the proton-neutron hypothesis, see Burcham, 1963, sec. 9-1.
[2] Burcham, 1963, chap. 4.

Positron   Positively charged electron of mass m_0

Photon   Quantum of electromagnetic radiation, commonly apparent as light, x ray, or gamma ray

A given nuclide is specified by a symbol like Li^7, $_3Li^7$, or $_3Li_4^7$. The letters denote the element. The right superscript gives the mass number A. The left subscript gives the atomic number Z, the right subscript the neutron number N. By recent convention the mass number is often given as the left superscript, making the symbol ^7Li, $_3^7Li$, or $_3^7Li_4$. In this book a nucleus in an excited state is denoted by the symbol with a right superscript star, e.g., Li^{7*}.

PROBLEMS

1-1 (a) An alpha particle of kinetic energy T_α makes a head-on collision with a nucleus of atomic number Z and mass number A. Calculate the distance of closest approach, taking into account the recoil of the nucleus. (b) An 0.2-Mev proton makes a head-on collision with an alpha particle at rest. What is the distance of closest approach (in F)? (c) If an alpha particle makes a head-on collision with a proton at rest, what must be its kinetic energy so that the distance of closest approach is identical to case (b)?

1-2 (a) A nucleus of mass number A makes a transition from an excited state to the ground state by emission of a gamma ray. What is the difference between the excitation energy E and the gamma-ray energy E_r due to the fact that the nucleus recoils? [The momentum of a photon is given by $p_r = E_r/c$. See Eqs. (2-1) and (2-3).] (b) If the above gamma ray is absorbed by a second nucleus of mass number A, to what energy can it excite the second nucleus? (c) Apply your results to the case of the Fe^{57} nucleus which emits a 14-kev gamma ray.

1-3 If the radius of a nucleus is given by Eq. (1-5) with $R_0 = 1.2$ F, what is the density of the nuclear matter (a) in g/cm^3, (b) in $nucleons/F^3$?

1-4 Suppose that the density of nucleons ρ in a nucleus varies with a radial distance r from the center of the nucleus as shown in the figure below. What fraction of the nucleons lie in the surface region in the nuclei Al^{27}, Te^{125}, and Po^{216} if $\rho_0 = 0.17$ F^{-3}, $c = 1.1$ $A^{1/3}$ F, $a = 3.0$ F? (This problem can be solved without evaluating any complicated integrals.)

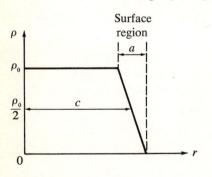

1-5 For the discussion of certain nuclear properties, it is convenient to define a root-mean-square radius by

$$R_{\text{rms}} = \left(\int_0^\infty \rho r^4 \, dr \Big/ \int_0^\infty \rho r^2 \, dr \right)^{\frac{1}{2}}$$

(a) Evaluate this for a nucleus of uniform density and for a nucleus whose density is given by the figure of Prob. 1.4. (b) Apply your result to the nucleus Te^{125}.

1-6 In Appendix A-1, it is shown that the density of nucleons in a deuterium nucleus has a radial dependence approximately given by $\rho = \rho_0 r^{-2} e^{-2\kappa r}$, where $\kappa^{-1} = 4.3$ F. (a) Evaluate the rms radius of the deuteron, using the definition given in Prob. 1-5. (b) Evaluate ρ_0.

NUCLEAR STRUCTURE ▨▨▨▨▨▨▨▨▨▨▨▨▨ 2

2-1 INTRODUCTION

Before discussing the structure of nuclei, we point out certain similarities and differences with the electronic structure of atoms. Atomic electrons are arranged in orbits, more accurately termed energy states, subject to the laws of quantum mechanics. In each atom the electrons are distributed over several states as a result of the so-called *Pauli exclusion principle*. Atomic electrons can be excited into normally unoccupied states, or can be removed completely from the atom. From such phenomena we are able to deduce the electronic structure of atoms.

In nuclei there are two groups of like particles: protons, and neutrons. Evidence will be presented showing that each of these groups is separately distributed over certain energy states subject to the restrictions of the Pauli exclusion principle.[1] Nuclei have excited states, and nucleons can be removed from, or added to, nuclei. Much information about nuclear structure can be obtained from a study of these phenomena.

Electrons and nucleons have intrinsic angular momenta, called *intrinsic spins*. The total angular momentum of a system of interacting particles reflects details of the forces between the particles. For example, from the (vector) addition, or coupling, of electron angular momenta in atoms we can infer the existence of a force connecting the spin and the orbital motion of an electron in the electric field of the nucleus (*spin-orbit coupling*). In nuclei, there is also a correlation between the orbital motion of each nucleon and its intrinsic spin, but not of the same origin as for an atomic electron. In addition, the nuclear force between two nucleons depends strongly on the relative orientation of their spins.

The structure of nuclei is more complex than that of atoms. In an atom the nucleus provides a common center of attraction for the electrons, whereas interelectronic forces generally play a secondary role. Furthermore, the predominant (coulomb) force is well understood. In nuclei there is no center of attraction; the nucleons are held together by their mutual interactions which turn out to be very complicated in detail. Nevertheless, as we will see, the short range of nuclear forces and the Pauli exclusion principle conspire to provide an effective overall force center for each nucleon. Also, atomic electrons represent one group of like particles, whereas in nuclei there are two different groups of like particles. This allows a richer variety of structures: there are about 100 types of atoms, but to date well over 1,000 different nuclides have been found.

2-2 ELEMENTS OF QUANTUM MECHANICS[2]

Neither atomic nor nuclear structure can be understood without the concepts of quantum mechanics. The continuity of presentation will, therefore, be interrupted to present some consequences of Schrödinger's equation which replaces the classical equation of motion. The solutions of this equation for two simple situations illustrate the physical basis for many aspects of nuclear structure.

Because the main idea of quantum mechanics is brought out by the de Broglie wave, a brief review of this concept will be given first.

2-2a de Broglie waves.   In the years 1900 to 1930 several decisive experiments were performed which demonstrated that classical mechanics, based on Newton's law of motion, and classical electromagnetism, based on Maxwell's equations,

[1] See Sec. 2-5.
[2] The reader who is familiar with quantum mechanics may wish to proceed directly to the end of this section.

fail to describe the behavior of atomic and subatomic particles. For example, experiments on the emission and absorption of electromagnetic radiation showed that the energy E_r of the radiation could be emitted (Planck, 1901) and absorbed (Einstein, 1905) only in bundles of energy, called *quanta*, rather than in a continuous manner as implied by Maxwell's equations for the electromagnetic field. Each quantum has the value

$$E_r = h\nu \qquad (2\text{-}1)$$

where h = Planck's constant ($h = 6.62 \times 10^{-27}$ erg-sec)

ν = frequency of electromagnetic radiation

The quantum energy is conveniently related to the wavelength λ of electromagnetic radiation by the relation

$$E_r \text{ (in Mev)} = \frac{1{,}240}{\lambda \text{ (in F)}} \qquad (2\text{-}2)$$

The scattering of x rays by atomic electrons (Compton, 1923) provided evidence that the linear momentum p_r of each quantum of electromagnetic radiation is given by

$$p_r = \frac{h}{\lambda} \qquad (2\text{-}3)$$

Therefore it is convenient to think of electromagnetic radiation as consisting of photons with particlelike mechanical properties.

De Broglie (1924) proposed that, conversely, particles should have wavelike properties. Assuming that this *de Broglie wave* is sinusoidal, the frequency and wavelength were to be given by the inverse relations to (2-1) and (2-3)

$$\nu_d = \frac{W}{h} \qquad (2\text{-}4)$$

$$\lambda_d = \frac{h}{p} \qquad (2\text{-}5)$$

where W is identified with the relativistic total energy[1] of the particle

$$W = mc^2 \qquad (2\text{-}6)$$

$$m = \frac{m_0}{(1 - v^2/c^2)^{\frac{1}{2}}} \qquad (2\text{-}7)$$

and p is the linear momentum

$$p = mv \qquad (2\text{-}8)$$

---

[1] For an elementary treatment of relativistic effects in mechanics, see Kittell, Knight, and Ruderman, 1965, vol. 1, chaps. 10–12. The phrase "total energy" is used with two different meanings in this book. The relativistic total energy is the sum of the rest energy, the kinetic energy, and the potential energy (if present). The nonrelativistic total energy is the sum of the kinetic and the potential energies.

In these relations

$$m = \text{total mass of particle}$$
$$m_0 = \text{rest mass of particle}$$
$$v = \text{speed of particle[1]}$$
$$c = \text{speed of light}$$

For later discussions we note that

$$W = m_0 c^2 + T \tag{2-9}$$

where T is the kinetic energy of the particle. For $v \ll c$, $T \approx \frac{1}{2}m_0 v^2 = \frac{1}{2}p^2/m_0$. Also, we can show from Eqs. (2-6) to (2-8) that

$$W^2 = p^2 c^2 + m_0^2 c^4 \tag{2-10}$$

which, for $m_0 = 0$, includes the relation between E_r and p_r obtained from Eqs. (2-1) and (2-3). For a neutron or proton, we find from Eq. (2-5) for $v \ll c$ that the de Broglie wavelength is related to the kinetic energy T by

$$\lambda_d \text{ (in F)} = \frac{28.6}{[T \text{ (in Mev)}]^{\frac{1}{2}}} \tag{2-11}$$

Electron scattering experiments on nickel crystals (Davisson and Germer, 1927) gave conclusive evidence that the de Broglie hypothesis (2-5) indeed has a basis in reality; but already in 1926 Schrödinger had proposed a differential equation for more general de Broglie waves than sinusoidal ones.

With the Planck hypothesis (2-1) and an essentially ad hoc assumption that the *orbital* angular momentum L of atomic electrons is quantized[2]

$$L = \text{integer} \times \hbar \tag{2-12}$$

Bohr (1913) proposed a model for the hydrogen atom which explained the optical emission spectrum of atomic hydrogen. The Schrödinger equation allows a reinterpretation of Bohr's model and provides a satisfactory basis for relation (2-12).

---

[1] In this book we will distinguish carefully between velocity **v**, which is a vector quantity, and speed v, which is the magnitude of velocity.

[2] In classical mechanics, the orbital angular momentum of a particle of mass m_0 moving about a fixed origin at a constant radial distance r with a speed v is equal to $m_0 vr$. More generally, the orbital angular momentum is a vector, given by

$$\mathbf{L} = \mathbf{r} \times m\mathbf{v} = \mathbf{r} \times \mathbf{p} \tag{2-13}$$

for any radius vector **r** and velocity **v**. For a system of particles, the total orbital angular momentum is the *vector sum* of the individual momenta. It can be shown for a system not subject to an external torque (if $v \ll c$) that the total orbital angular momentum either about a fixed origin or about the center of mass of the system is a constant vector, i.e., constant in length, direction, and sense.

2-2b Schrödinger equation.   The de Broglie wave can be considered as a mathematical wave *guiding* the motion of a particle (Born, 1926). Its amplitude Ψ is a function of space and time. Although it is not possible to derive the equation governing the wave, various plausible arguments can be used to relate it to familiar concepts such as conservation of energy. In this connection, we note that in classical mechanics Newton's laws of motion also cannot be derived, but are descriptions of experimental facts.

Schrödinger's equation is

$$-\frac{\hbar^2}{2m_0}\nabla^2\Psi + V\Psi = i\hbar\frac{\partial\Psi}{\partial t} \tag{2-14}$$

where, in Cartesian coordinates,

$$\Psi = \Psi(x,y,z,t) = \text{\textit{wave function}}\ \text{of particle}$$

$$\nabla^2\Psi = \frac{\partial^2\Psi}{\partial x^2} + \frac{\partial^2\Psi}{\partial y^2} + \frac{\partial^2\Psi}{\partial z^2} \tag{2-15}$$

$$V = V(x,y,z,t) = \text{potential energy of particle[1]}$$

and $\qquad\qquad i = \sqrt{-1}$

If V is independent of the time, we can separate space and time variables by setting[2]

$$\Psi = \psi(x,y,z)\tau(t) \tag{2-16}$$

Substituting into Eq. (2-14) and dividing by $\psi\tau$ we find

$$-\frac{\hbar^2}{2m_0}\frac{\nabla^2\psi}{\psi} + V = \frac{i\hbar}{\tau}\frac{d\tau}{dt} \tag{2-17}$$

Since the left-hand side of this equation depends only on space variables and the right-hand side only on the time, the equation cannot be satisfied for all points in space at all times unless each side has the same *constant* value. Call this constant E. (We will see below that this is the nonrelativistic total energy of the system.) From the right-hand side of Eq. (2-17) we then obtain

$$\tau = Ce^{-i(E/\hbar)t} \tag{2-18}$$

For convenience we will set the arbitrary constant C equal to unity. The left-hand side of Eq. (2-17) can be written

$$-\frac{\hbar^2}{2m_0}\nabla^2\psi + V\psi = E\psi \tag{2-19}$$

---

[1] As in classical mechanics, the potential energy is defined only within an arbitrary constant. For the purposes of this book we will usually define V so that it is zero for $x, y, z \to \infty$.
[2] This is a general technique for solving certain types of partial differential equations.

This is called the *time-independent Schrödinger equation*. It is an equation to which we will refer often.

To interpret the equation, assume that the wave function ψ depends only on one coordinate, say x

$$\psi = \psi(x) \tag{2-20}$$

This can occur only if V is a function of x alone. Then,

$$\nabla^2\psi = \frac{d^2\psi}{dx^2} \tag{2-21}$$

and Eq. (2-19) can be put into the form

$$\frac{d^2\psi}{dx^2} = -k^2\psi \tag{2-22}$$

where k (called *wave number*) is defined by

$$\frac{\hbar^2 k^2}{2m_0} = E - V(x) \tag{2-23}$$

If V happens to be *independent* of x so that k is also independent of x, Eq. (2-22) is mathematically equivalent to a simple-harmonic-oscillator equation with the solution

$$\psi = ae^{ikx} + be^{-ikx} \tag{2-24}$$

The arbitrary constants a and b are determined by boundary conditions. One special form of Eq. (2-24) is the sine function

$$\psi = A \sin kx \tag{2-25}$$

which is the de Broglie wave mentioned in Sec. 2-2a. The wave number k is therefore related to the de Broglie wavelength by

$$k = \frac{2\pi}{\lambda_d} \tag{2-26}$$

and hence to the momentum p of the particle by

$$k = \frac{p}{\hbar} \tag{2-27}$$

according to Eq. (2-5). The plausibility of Eq. (2-27) is enhanced if we substitute it into Eq. (2-23) to find the classical law of conservation of energy

$$\frac{p^2}{2m_0} + V = E \tag{2-28}$$

We note that this law is obtained in classical mechanics if the forces acting on the particle are conservative or, in other words, if the particle has a potential energy which does not depend directly on time. Exactly the same assumption was made in expression (2-16) from which all subsequent equations followed. We can therefore think of Eq. (2-19) as the quantum mechanical equivalent of the law of conservation of energy.

We can go one step further and note from Eq. (2-18) that the angular frequency ω ($= 2\pi\nu$) of the (oscillating) wave function is given by

$$\omega = \frac{E}{\hbar} \qquad (2\text{-}29)$$

This corresponds to the de Broglie relation (2-4). The reason E rather than W occurs in Eq. (2-28) is that the Schrödinger equation is the quantum mechanical analog of the nonrelativistic energy law. The quantum mechanical replacement for the relativistic energy expression is also known and was discovered by Dirac (1928). Some of its consequences will be mentioned in Sec. 3-4d.

2-2c Interpretation of Ψ. Boundary conditions. The simple solution (2-24) can serve to illustrate the two types of waves which Schrödinger's equation (2-14) can generate when V is independent of t: *standing waves* and *traveling waves*. To obtain the former, we write Eq. (2-16) with the help of (2-18) and (2-29) in the form

$$\Psi = (ae^{ikx} + be^{-ikx})e^{-i\omega t} \qquad (2\text{-}30)$$

This is reminiscent of a standing wave on a string, whose deflection y from equilibrium is given by

$$y = (a \sin kx + b \cos kx) \sin \omega t \qquad (2\text{-}31)$$

To obtain traveling waves, we write

$$\Psi = ae^{i(kx-\omega t)} + be^{-i(kx+\omega t)} \qquad (2\text{-}32)$$

which recalls a sine wave traveling in the $+x$ direction

$$y = a \sin (kx - \omega t) \qquad (2\text{-}33)$$

and a wave traveling in the $-x$ direction

$$y = b \sin (kx + \omega t) \qquad (2\text{-}34)$$

For future purposes it is useful to note that we can always recognize from the sign of E whether a general wave function Ψ will be a standing or traveling wave. To show this, assume a one-dimensional situation and let the forces on the particle extend only over a finite region of space so that

$$V(x) \rightarrow 0 \qquad \text{as} \qquad |x| \rightarrow \infty \qquad (2\text{-}35)$$

From Eq. (2-23) it follows that as $|x| \rightarrow \infty$

$$k \rightarrow \left(\frac{2m_0 E}{\hbar^2} \right)^{\frac{1}{2}} \qquad (2\text{-}36)$$

If $E > 0$, k will be a real quantity. The solutions of Eq. (2-22) can then be written in the form (2-32) at large distance from the origin, so that Ψ is a traveling wave solution. If $E < 0$, Eq. (2-36) shows that k is pure imaginary far from the origin, that is $k \rightarrow i\kappa$, where κ is a real quantity. From Eq. (2-24) it then follows that

$$\psi \rightarrow ae^{-\kappa x} + be^{+\kappa x} \tag{2-37}$$

where
$$\kappa = \left(\frac{2m_0 |E|}{\hbar^2}\right)^{\frac{1}{2}} \tag{2-38}$$

Substituting into Eq. (2-16)

$$\Psi \rightarrow \psi(x)e^{-i\omega t} \tag{2-39}$$

where $\psi(x)$ is a real function. This has the mathematical form of a standing wave. Therefore, positive E means that at a large distance from the force center Ψ is a traveling wave and negative E means that it is a standing wave.

In accordance with the physical conditions on Ψ, given below, we must set $b = 0$ or $a = 0$ in Eq. (2-37) as $x \rightarrow +\infty$ or $-\infty$ respectively. Since far from the force center Ψ becomes infinitesimally small, Eq. (2-39) tells us that a standing wave solution represents a *localized* oscillating disturbance. The corresponding state of the particle is therefore called a *bound state*. A traveling wave, on the other hand, represents a disturbance coming towards a force center or traveling away from it, such as occurs in the scattering of a particle or the diffraction of a light wave by a slit.

After considering various scattering processes on the basis of Schrödinger's equation, Born (1926) suggested that $\Psi(x,y,z,t)$ should be regarded as a "ghost wave" guiding the motion of a particle. He found that the expression[1]

$$\Psi^*\Psi \, dx \, dy \, dz = |\Psi|^2 \, dx \, dy \, dz \tag{2-40}$$

represents the probability of finding the particle in a particular volume element $dx \, dy \, dz$. The quantity $|\Psi(x,y,z,t)|^2$, itself, is the probability per unit volume of finding the particle near the point x, y, z. A connection with the classical probability will be made in Sec. 2-2f.

From Eq. (2-16) and from the form (2-18) of τ, we see that for time-independent potentials

$$|\Psi(x,y,z,t)|^2 = |\psi(x,y,z)|^2 \tag{2-41}$$

Hence $|\Psi|^2$ is independent of time in that case.

Since $|\Psi|^2$ is a physical probability, Ψ has the following properties:

1 It must be single valued and continuous everywhere.

2 All first partial derivatives of Ψ, which turn out to be related to the current density or flux of particles (number of particles per unit time per unit cross-sectional area), must also be continuous.

---

[1] A superscript star denotes the complex conjugate of the quantity in question. Bars denote the absolute magnitude. Note that $(a + ib)^*(a + ib) = (a - ib)(a + ib) = a^2 + b^2 = |a + ib|^2$.

3 Ψ must never be infinite.

4 If $V \to \infty$, $\Psi \to 0$, since each term in Eq. (2-19) must remain finite.

Furthermore, for one particle localized in a certain region of space, such as an atomic electron in a H atom,

$$\int_{\text{all space}} |\psi|^2 \, dx \, dy \, dz = 1 \tag{2-42}$$

so that as $r \to \infty$, $\psi \to 0$ for any bound state. Equation (2-42) is called the *normalization condition*.

FIGURE 2-1 Spherical coordinates. (*a*) One-particle system. (*b*) Two-particle system. The center of mass is denoted by C.

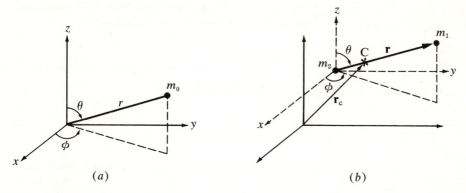

(*a*)   (*b*)

2-2d Schrödinger equation in spherical coordinates.   Many physical potentials, such as the coulomb potential, have spherical symmetry. In this case, one can show[1] that the general wave function can be separated in spherical coordinates r, θ, and ϕ, illustrated in Fig. 2-1*a*.

$$\psi(r,\theta,\phi) = R(r)\Theta(\theta)\Phi(\phi) \tag{2-43}$$

Imposition of the conditions that ψ must be single-valued and finite everywhere restricts Θ and Φ to the form

$$\Theta(\theta) = P_l^{(m)}(\cos \theta) \tag{2-44}$$

$$\Phi(\phi) = e^{im\phi} \tag{2-45}$$

where $P_l^{(m)}$ = associated Legendre polynomial[2] of the order l in $\cos \theta$

l, m = integers (including zero) with $|m| \leq l$, and m positive or negative, but l positive only

---

[1] Schiff, 1955, "Quantum Mechanics," sec. 14. See also Prob. 2-7 at the end of this chapter.
[2] The lowest-order associated Legendre polynomials are $P_0^{(0)} = 1$; $P_1^{(0)} = \cos \theta$; $P_1^{(\pm 1)} = (1 - \cos^2 \theta)^{\frac{1}{2}} = \sin \theta$.

One can show that the integers l and m are related to the *orbital* angular momentum **L** of the particle (about the origin). The magnitude of **L** is

$$L = [l(l+1)]^{\frac{1}{2}}\hbar \qquad (2\text{-}46)$$

and its z component is equal to $m\hbar$.

The *radial function* $R(r)$ in Eq. (2-43) is given by

$$-\frac{\hbar^2}{2m_0}\frac{d^2u}{dr^2} + \left[\frac{l(l+1)\hbar^2}{2m_0r^2} + V(r)\right]u = Eu \qquad (2\text{-}47)$$

where the substitution

$$u = rR(r) \qquad (2\text{-}48)$$

has been made to simplify the equation. The general form of the equation can be appreciated on the basis of classical concepts. Consider the motion of a particle in a central force field.[1] The particle will move in a plane. Decompose

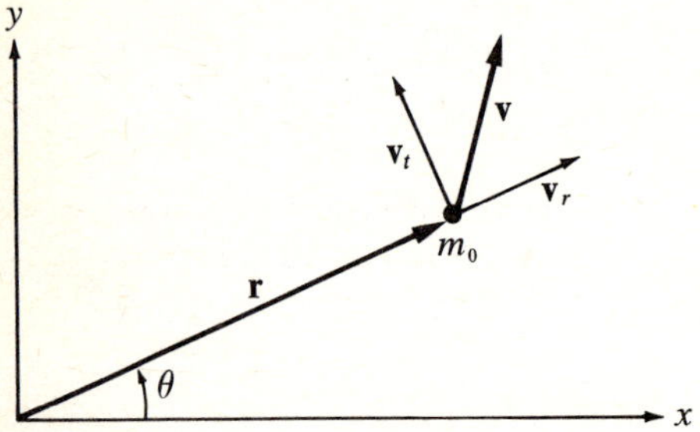

FIGURE 2-2 Classical motion of a particle in a central force field. The plane shown is the plane in which the trajectory of the particle lies.

its instantaneous velocity **v** into components $\mathbf{v}_r$ and $\mathbf{v}_t$ which, respectively, are radial and tangential to the instantaneous radius vector **r** of the particle, as shown in Fig. 2-2. Conservation of energy gives

$$\tfrac{1}{2}m_0(v_r^2 + v_t^2) + V(r,\theta) = E \qquad (2\text{-}49)$$

where $V(r,\theta)$ is the potential associated with the central force field. Since there can be no torque exerted on the particle by a central force, the orbital angular momentum L of the particle

$$L = m_0 v_t r \qquad (2\text{-}50)$$

is a constant of the motion. Eliminating v_t between Eqs. (2-49) and (2-50) we obtain

$$\tfrac{1}{2}m_0 v_r^2 + \frac{L^2}{2m_0r^2} + V(r,\theta) = E \qquad (2\text{-}51)$$

This equation has the same relation to Eq. (2-47) as Eq. (2-28) has to Eq. (2-19).

[1] In classical mechanics, the spherically symmetric type of potential considered in this section would be a special case of a central force field.

We see that the transition from classical to quantum mechanics makes plausible the substitution

$$L^2 \rightarrow l(l+1)\hbar^2 \tag{2-52}$$

in agreement with relation (2-46) which can be derived directly. The substitution (2-52) will be used several times in this book.

Some characteristics of the radial solutions [Eq. (2-47)] will be discussed in Sec. 2-5b. Here it is of interest only to point out that the equation for the case $l = 0$ (or $L = 0$, i.e., zero orbital angular momentum)

$$-\frac{\hbar^2}{2m_0}\frac{d^2u}{dr^2} + V(r)u = Eu \tag{2-53}$$

has a *mathematical* form which is identical to the one-dimensional equation (2-22). This identity will be helpful later on. We should note, though, that the definition (2-48) of u requires always

$$u = 0 \quad \text{at} \quad r = 0 \tag{2-54}$$

because $R(r)$ must remain finite everywhere. (See condition *3* of the boundary conditions in Sec. 2-2c.)

2-2e Wave equation for two particles under mutual forces. In nuclear physics problems, the motion of two particles subject only to mutual forces is very common. It is, therefore, worthwhile to compare the classical method of separating the motion *of* the center of mass and the motion *about* the center of mass with the quantum mechanical separation. If two particles of mass m_1 and m_2 move under mutual forces $\mathbf{F}_1$ and $\mathbf{F}_2$, the classical equations of motion of either particle with respect to a fixed origin are

$$\mathbf{F}_1 = m_1 \frac{d^2\mathbf{r}_1}{dt^2}$$
$$\tag{2-55}$$
$$\mathbf{F}_2 = m_2 \frac{d^2\mathbf{r}_2}{dt^2}$$

where
$$\mathbf{F}_1 = -\mathbf{F}_2 \tag{2-56}$$

Defining the coordinate of the center of mass by

$$\mathbf{r}_c = \frac{\mathbf{r}_1 m_1 + \mathbf{r}_2 m_2}{m_1 + m_2} \tag{2-57}$$

the motion of each particle can be expressed with respect to the center of mass of the system. Under condition (2-56) the total force $\mathbf{F}_1 + \mathbf{F}_2$ on the system $m_1 + m_2$ is zero and therefore the center of mass moves with constant (vector) velocity. With respect to the center of mass, particle 1 has the radius vector

$$\mathbf{r}\frac{m_2}{m_1 + m_2} \tag{2-58}$$

and particle 2 the radius vector $-rm_1/(m_1 + m_2)$. Here

$$\mathbf{r} = \mathbf{r}_1 - \mathbf{r}_2 \tag{2-59}$$

refers to the relative separation of the particles. The equation of motion of $\mathbf{r}$ can be derived by substituting in Eq. (2-55) for $\mathbf{r}_1$ the expression

$$\mathbf{r}_1 = \mathbf{r}_c + \mathbf{r}\frac{m_2}{m_1 + m_2} \tag{2-60}$$

Since $d\mathbf{r}_c/dt = \text{constant}$, we find

$$\mathbf{F}_1 = M_0 \frac{d^2\mathbf{r}}{dt^2} \tag{2-61}$$

where

$$M_0 = \frac{m_1 m_2}{m_1 + m_2} \tag{2-62}$$

is the reduced mass of the system.

A separation of the motion similar to the classical one can be made for Schrödinger's equation. For this purpose we note that for two particles Eq. (2-19) becomes

$$-\frac{\hbar^2}{2m_1} \nabla_1^2\psi - \frac{\hbar^2}{2m_2} \nabla_2^2\psi + V\psi = E\psi \tag{2-63}$$

The wave function ψ depends on $\mathbf{r}_1$ and $\mathbf{r}_2$, but for mutual forces V depends only on $\mathbf{r}_1 - \mathbf{r}_2 = \mathbf{r}$. Using Cartesian coordinates, $\nabla_1^2\psi$ represents

$$\nabla_1^2\psi = \frac{\partial^2\psi}{\partial x_1^2} + \frac{\partial^2\psi}{\partial y_1^2} + \frac{\partial^2\psi}{\partial z_1^2} \tag{2-64}$$

and similarly for $\nabla_2^2\psi$. Since x_1 and x_2 are functions of x_c and x as given by relations (2-57) and (2-59)

$$\frac{\partial\psi}{\partial x_1} = \frac{\partial\psi}{\partial x}\frac{\partial x}{\partial x_1} + \frac{\partial\psi}{\partial x_c}\frac{\partial x_c}{\partial x_1}$$

$$= \frac{\partial\psi}{\partial x} + \frac{\partial\psi}{\partial x_c}\frac{m_1}{m_1 + m_2} \tag{2-65}$$

and

$$\frac{\partial^2\psi}{\partial x_1^2} = \frac{\partial}{\partial x_1}\left(\frac{\partial\psi}{\partial x_1}\right) = \frac{\partial}{\partial x}\left(\frac{\partial\psi}{\partial x_1}\right)\frac{\partial x}{\partial x_1} + \frac{\partial}{\partial x_c}\left(\frac{\partial\psi}{\partial x_1}\right)\frac{\partial x_c}{\partial x_1}$$

$$= \frac{\partial^2\psi}{\partial x^2} + \frac{\partial^2\psi}{\partial x\,\partial x_c}\frac{2m_1}{m_1 + m_2} + \frac{\partial^2\psi}{\partial x_c^2}\left(\frac{m_1}{m_1 + m_2}\right)^2 \tag{2-66}$$

Similarly

$$\frac{\partial^2\psi}{\partial x_2^2} = \frac{\partial^2\psi}{\partial x^2} - \frac{\partial^2\psi}{\partial x\,\partial x_c}\frac{2m_2}{m_1 + m_2} + \frac{\partial^2\psi}{\partial x_c^2}\left(\frac{m_2}{m_1 + m_2}\right)^2 \tag{2-67}$$

Eq. (2-63) can now be written

$$-\frac{\hbar^2}{2M_0}\left(\frac{\partial^2\psi}{\partial x^2}+\frac{\partial^2\psi}{\partial y^2}+\frac{\partial^2\psi}{\partial z^2}\right)-\frac{\hbar^2}{2(m_1+m_2)}\left(\frac{\partial^2\psi}{\partial x_c^2}+\frac{\partial^2\psi}{\partial y_c^2}+\frac{\partial^2\psi}{\partial z_c^2}\right)+V\psi=E\psi$$

(2-68)

where M_0 is the reduced mass defined by Eq. (2-62). Since V depends only on x, y, z, we can separate variables in this equation by the substitution

$$\psi(\mathbf{r},\mathbf{r}_c)=\psi_0(\mathbf{r})\psi_c(\mathbf{r}_c)$$

(2-69)

This yields, after division by $\psi_0\psi_c$,

$$\left[-\frac{\hbar^2}{2M_0}\frac{\nabla^2\psi_0}{\psi_0}+V\right]+\left[-\frac{\hbar^2}{2(m_1+m_2)}\frac{\nabla^2\psi_c}{\psi_c}\right]=E$$

(2-70)

Dividing the total energy E into the energy E_c of the center-of-mass (c.m.) motion and the energy E_0 with respect to the center of mass, the separate coordinate dependence of the two bracketed quantities requires

$$-\frac{\hbar^2}{2M_0}\nabla^2\psi_0+V\psi_0=E_0\psi_0$$

(2-71)

and

$$-\frac{\hbar^2}{2(m_1+m_2)}\nabla^2\psi_c=E_c\psi_c$$

(2-72)

The first of these equations resembles in all respects Eq. (2-19) for a single particle, and is the quantum-mechanical analog of the energy equation corresponding to Eq. (2-61). The second equation represents the c.m. motion with constant velocity; this can be seen from the discussion following Eq. (2-23) with V set equal to zero.

In future expressions for a two-particle system, the subscript zero will be dropped. We should always remember, though, that in the radial equation (2-47), corresponding to Eq. (2-71), the orbital angular momentum expression $[l(l+1)]^{\frac{1}{2}}\hbar$ now refers to the sum of the orbital angular momenta of both particles about the center of mass.

2-2f Particle in a closed cubical box. We now apply the Schrödinger equation (2-19) to two simple problems which give us some insight into the properties of a quantum mechanical system such as a nucleus. The first of these, a particle in a closed box, simulates a situation in which a particle is in a bound state, such as an electron in an atom or a nucleon in a nucleus. The second problem is concerned with a beam of particles.

A closed box must be represented by a potential which is infinitely high at the position of the walls, since the particle cannot be outside the box and hence ψ must be zero everywhere outside the box (see condition 4 of the boundary

FIGURE 2-3   Particle in a closed cubical box.   (*a*) Location of box.   (*b*) Potential shape. *E* is the total energy of the particle.   (*c*) Shape of a typical wave function along *x* direction ($n_x = 4$).   (*d*) Shape of the probability density corresponding to (*c*). The classical probability density is indicated by a dashed line.

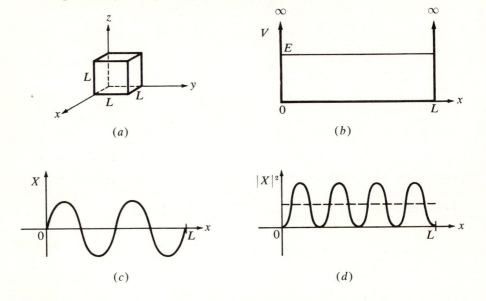

conditions in Sec. 2-2c). For convenience we can put $V = 0$ inside the box,[1] as indicated in Fig. 2-3*b*. Substituting this into Eq. (2-19) and setting

$$\psi(x,y,z) = X(x)Y(y)Z(z) \tag{2-73}$$

it is easy to separate variables and to obtain

$$\frac{1}{X}\frac{d^2x}{dx^2} + \frac{1}{Y}\frac{d^2Y}{dy^2} + \frac{1}{Z}\frac{d^2Z}{dz^2} = -k^2 \tag{2-74}$$

where k^2 is defined in terms of the (constant) energy E and mass m_0 of the particle by

$$k^2 = \frac{2m_0E}{\hbar^2} \tag{2-75}$$

Each of the three terms on the left-hand side of Eq. (2-74) depends on a different, independent, coordinate. Since the sum of the three terms is equal to a constant, each of the terms must be equal to a constant. For example, we can set

$$\frac{1}{X}\frac{d^2X}{dx^2} = -k_x^2 \tag{2-76}$$

[1] There is no particular meaning attached to $V = 0$. See footnote following Eq. (2-15).

and similarly for the other two equations, so that the separate constants are related by

$$k_x{}^2 + k_y{}^2 + k_z{}^2 = k^2 \tag{2-77}$$

Equation (2-76) is identical to Eq. (2-22) and its solution is

$$X = a_x e^{ik_x x} + b_x e^{-ik_x x} \tag{2-78}$$

The arbitrary constants a_x and b_x are determined by the boundary condition 4 of Sec. 2-2c:

$$X(0) = 0 \qquad X(L) = 0 \tag{2-79}$$

Recalling that $e^{is} = \cos s + i \sin s$, we see that the first of these conditions requires

$$X = A_x \sin k_x x \tag{2-80}$$

where A_x is an arbitrary constant, and the second condition restricts k_x to the values

$$k_x = \frac{n_x \pi}{L} \tag{2-81}$$

with n_x equal to an integer. It is important to note for later work that only positive integers are of interest. A change of sign of n_x is equivalent to a sign change of the arbitrary constant A_x in Eq. (2-80) and therefore does not produce a new wave function. The value $n_x = 0$ is excluded, because in that case $\psi = 0$ throughout the box; Eq. (2-42) shows that there could then be no particle in the box.

Proceeding similarly for the y and z solutions we find

$$Y = A_y \sin k_y y \qquad Z = A_z \sin k_z z \tag{2-82}$$

where

$$k_y = \frac{n_y \pi}{L} \quad \text{and} \quad k_z = \frac{n_z \pi}{L} \tag{2-83}$$

Summarizing Eqs. (2-80) and (2-82), the complete wave function is

$$\psi = A \sin k_x x \sin k_y y \sin k_z z \tag{2-84}$$

where

$$A = A_x A_y A_z \tag{2-85}$$

Each component of this function is a simple de Broglie wave. For example, the wavelength in the x direction is equal to

$$\lambda_x = \frac{2\pi}{k_x} = \frac{2L}{n_x} \tag{2-86}$$

This is just the condition for producing a standing wave in the box with nodes at the walls.

The constant A in Eq. (2-84) can be determined by the normalization condition (2-42).

$$1 = A^2 \int_0^L \sin^2 \frac{n_x \pi x}{L} \, dx \int_0^L \sin^2 \frac{n_y \pi y}{L} \, dy \int_0^L \sin^2 \frac{n_z \pi z}{L} \, dz$$

$$= A^2 (\tfrac{1}{2} L)^3 \tag{2-87}$$

(Each of the integrals is equal to $L/2$ by inspection, since the average value of the sine-square function is equal to $\frac{1}{2}$ over any number of complete half-periods.) Choosing the positive sign for convenience,

$$A = (2/L)^{\frac{3}{2}} \tag{2-88}$$

yields the complete normalized solution

$$\psi = \left(\frac{2}{L}\right)^{\frac{3}{2}} \sin \frac{n_x \pi x}{L} \sin \frac{n_y \pi y}{L} \sin \frac{n_z \pi z}{L} \tag{2-89}$$

Let us compare the probability density $|\psi|^2$ with its classical value. From the point of view of Newtonian mechanics, we are treating the problem of a particle bouncing around inside a closed box in a perfectly elastic manner, free from any external forces. Such a particle has a constant speed wherever it is within the box and, therefore, its probability density is equal to $1/L^3$. But this is exactly the value toward which $|\psi|^2$ will tend if $n_x, n_y, n_z \rightarrow \infty$, for then each sine-square function can be replaced by its average value $\frac{1}{2}$ (see Fig. 2-3c). This is an illustration of the *correspondence principle* (Bohr, 1923) according to which quantum mechanics can be approximated by classical mechanics whenever the quantum numbers of the system, here n_x, n_y, and n_z, become very large.

The boundary conditions (2-79) and similar equations for Y and Z restrict not only the form of the wave function but also the energy E of the system. Substitution of Eqs. (2-81) and (2-83) into (2-77) and (2-75) gives

$$E = (n_x^2 + n_y^2 + n_z^2) \frac{\pi^2 \hbar^2}{2m_0 L^2} \tag{2-90}$$

The boundary conditions on ψ, therefore, cause the quantization of the energy. Similarly, the finiteness and single-valuedness of ψ (conditions *1* and *3* of Sec. 2-2c) cause the quantization of the orbital angular momentum (Sec. 2-2d). Although this way of *explaining* quantization may seem unsatisfactory to the reader, it is not possible to give a better understanding by any analogies or concepts based on classical physics.

In Table 2-1 and Fig. 2-4, we show the lowest energy levels of the system in units of $\pi^2 \hbar^2/(2m_0 L^2)$. As mentioned above, no quantum number can have the value zero and only positive quantum numbers are considered. The lowest energy state, therefore, cannot be zero. We will show that this is in accord with the *uncertainty principle* (Heisenberg, 1927). In the lowest state, the x component

TABLE 2-1 Quantum numbers and energy levels of a particle in a closed cubical box

n_x	n_y	n_z	$(n_x^2 + n_y^2 + n_z^2)$†	Number of levels
1	1	1	3	1
1	1	2	6	3‡
1	2	2	9	3
1	1	3	11	3
2	2	2	12	1
1	2	3	14	6
2	2	3	17	3

† This is the energy E in units of $\pi^2\hbar^2/(2m_0L^2)$. See Eq. (2-90).
‡ The three levels correspond to $n_x = 2$, $n_y = 2$ and $n_z = 2$, respectively. All three levels have the same energy but different wave functions. They are called *degenerate*. A similar situation can occur for the other cases.

of linear momentum is uncertain by the approximate amount

$$\Delta p_x \approx 2p_x \qquad (2\text{-}91)$$

because the direction of travel of the particle cannot be determined from its wave function. The corresponding uncertainty in position is roughly equal to

$$\Delta x \approx L \qquad (2\text{-}92)$$

Since the wave function is just a half-sine wave with zeros at $x = 0$ and $x = L$,

FIGURE 2-4 Energy levels of a particle in a closed cubical box. Note that the lowest energy state is not at zero energy.

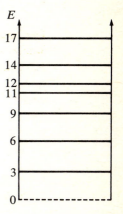

we cannot really locate the particle within a distance of the order of L. Recalling $p_x = \hbar k_x$ [Eq. (2-27)] and using Eq. (2-81), we obtain within a factor of 2

$$\Delta p_x \Delta x \approx h \qquad (2\text{-}93)$$

This is one expression of the uncertainty principle.

It is instructive to compute the magnitude of the characteristic energy step $\pi^2 \hbar^2 / (2m_0 L^2)$ which occurs in expression (2-90).

For an electron in an atom, $m_0 = 9.1 \times 10^{-28}$ g, $L \approx 10^{-8}$ cm

$$\frac{\pi^2 \hbar^2}{2m_0 L^2} \approx \frac{\pi^2 (1.05 \times 10^{-27})^2}{2 \times 9.1 \times 10^{-28} \times 10^{-16}}$$

$$\approx 0.5 \times 10^{-10} \text{ ergs} \approx 30 \text{ ev} \qquad (2\text{-}94)$$

For a nucleon in a typical nucleus, $m_0 = 1.6 \times 10^{-24}$ g, $L \approx 5 \times 10^{-13}$ cm

$$\frac{\pi^2 \hbar^2}{2m_0 L^2} \approx 6 \text{ Mev} \qquad (2\text{-}95)$$

These values are of the correct order of magnitude and allow us to appreciate the enormous difference between atomic and nuclear energies.

FIGURE 2-5 A simple potential barrier in one dimension. The particles originating at $x = -\infty$, and those traveling to the right and left are indicated by their wave functions with arrows. In each case, $\Psi = \psi e^{-i\omega t}$.

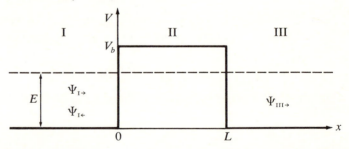

2-2g Barrier penetration of a particle.   The second problem to which we apply Schrödinger's equation (2-19) is the one-dimensional penetration of a potential barrier by a beam of particles. A classical analog of this situation would be a stream of marbles rolling normally up an incline. It is clear that if the (non-relativistic) total energy E of a marble does not exceed the maximum gravitational potential energy V_b corresponding to the top of the incline, the marble will always roll back, i.e., it will be *reflected by the potential barrier*. On the other hand, for $E > V_b$ the marble will always go over the top. In a system in which quantum effects are important, these results are strongly modified.

Figure 2-5 shows a simple potential barrier in one dimension. A stream of particles is supposed to originate at $x = -\infty$ and to travel toward the barrier, i.e., towards the right. Assume that each particle has a total energy E. For convenience divide space into three regions I, II, and III, and set $V_I = 0$, $V_{II} = V_b$, $V_{III} = 0$. Because $E > V$ for regions I and III, the particles in these regions are represented by the traveling wave solution (2-32) of Eq. (2-22). In region I, particles can be reflected by the barrier towards the left, but in region III, particles cannot travel towards the left because the source of particles is at $x = -\infty$ and nothing in region III can reflect the particles back to the left. The solutions of Eq. (2-22) in regions I and III are therefore

$$\psi_I = a_I e^{ikx} + b_I e^{-ikx} = \psi_{I\rightarrow} + \psi_{I\leftarrow} \tag{2-96}$$

$$\psi_{III} = a_{III} e^{ikx} = \psi_{III\rightarrow} \tag{2-97}$$

where $k^2 = 2m_0 E/\hbar^2$.

In region II the equation to be solved is

$$\frac{d^2\psi_{II}}{dx^2} = \kappa^2 \psi_{II} \tag{2-98}$$

where

$$\kappa^2 = 2m_0(V_b - E)/\hbar^2 \tag{2-99}$$

The solution is

$$\psi_{II} = a_{II} e^{\kappa x} + b_{II} e^{-\kappa x} \tag{2-100}$$

which is a standing-wave–type solution of the form (2-39).[1]

The probability P of transmission of the flux of particles through the barrier in the present case is equal to[2]

$$P = \frac{|\psi_{III\rightarrow}|^2 v}{|\psi_{I\rightarrow}|^2 v} = \frac{|a_{III}|^2}{|a_I|^2} \tag{2-101}$$

where v is the speed of the particles. To evaluate this expression we note that the (complex) coefficients a and b are determined by assuring that ψ and $d\psi/dx$ are continuous at $x = 0$ and L. For example, at $x = 0$,

$$\psi_{I\rightarrow} + \psi_{I\leftarrow} = \psi_{II} \qquad \text{leads to} \qquad a_I + b_I = a_{II} + b_{II} \tag{2-102}$$

and at $x = L$

$$\psi_{II} = \psi_{III\rightarrow} \qquad \text{leads to} \qquad a_{II} e^{\kappa L} + b_{II} e^{-\kappa L} = a_{III} e^{ikL} \tag{2-103}$$

Similar equations are obtained for the derivatives.

[1] From a classical-mechanics point of view, it is of course puzzling that particles can be within the barrier region II at all, because their kinetic energy would be negative there. However, an application of the uncertainty principle to this problem shows that if we wished to find out whether a particle is really localized within the barrier we would have to give it momentum sufficient to make its kinetic energy positive.

[2] In classical terms, a beam of particles traveling with a velocity v has a current density nv, where n is the number of particles per unit volume in the beam. Current density is equal to the number of particles traversing unit area perpendicular to v in unit time. Flux is equal to the number of particles traversing, in any direction, unit area in unit time. For a beam of particles, current density or flux can be used interchangeably, if we remember to place the unit area perpendicular to the velocity vector. The correct quantum mechanical expression for current density is given in Schiff, 1955, sec. 7. For plane waves, it reduces to $|\psi|^2 \, v$.

Solving the preceding equations for a_{III} and a_I we find after some algebra[1]

$$P = \left[1 + \frac{V_b{}^2}{4E(V_b - E)} \sinh^2 \kappa L\right]^{-1} \qquad (2\text{-}104)$$

which for $\kappa L \gg 1$, i.e., $\sinh^2 \kappa L \approx \frac{1}{4}e^{+2\kappa L}$, becomes

$$P \approx 16\frac{E}{V_b}\left(1 - \frac{E}{V_b}\right)e^{-2\kappa L} \qquad (2\text{-}105)$$

The important factor in most physical cases is the exponential. For example, for a 5-Mev proton and $V_b = 10$ Mev, $L = 10^{-12}$ cm, Eq. (2-99) gives

$$\kappa = \frac{[2 \times 1.6 \times 10^{-24} \times (10 - 5) \times 1.6 \times 10^{-6}]^{\frac{1}{2}}}{1.05 \times 10^{-27}}$$

$$\approx 5 \times 10^{12} \text{ cm}^{-1}$$

Therefore $e^{-2\kappa L} = e^{-10} = 0.5 \times 10^{-4}$

and $P = (16 \times 0.5 \times 0.5) \times 0.5 \times 10^{-4} = 2 \times 10^{-4}$

Usually the term in front of the exponential is ignored and we write

$$P \approx e^{-\gamma} \qquad (2\text{-}106)$$

where $\gamma = 2\kappa L = 2[2m_0(V_b - E)]^{\frac{1}{2}}L/\hbar$.

If V is not a constant, but varies with x, we can show[2] that the same expression for P holds approximately but with

$$\gamma = \frac{2}{\hbar}\int_{x_1}^{x_2} \{2m_0[V(x) - E]\}^{\frac{1}{2}}\,dx \qquad (2\text{-}107)$$

where x_1 and x_2 are the classical *turning points*, i.e., the points at which $E = V(x)$ as shown in Fig. 2-6.

FIGURE 2-6 A general potential barrier in one dimension. The classical turning points x_1 and x_2 are indicated.

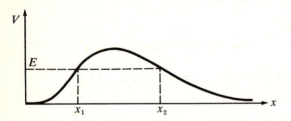

In Sec. 2-2d, we mentioned that for a problem with spherical symmetry and zero angular momentum, the same mathematical expressions are obtained as

---

[1] Evans, 1955, app. C.
[2] Schiff, 1955, sec. 28. Note that if the barrier is broken up between x_1 and x_2 into n adjacent barriers of thickness Δx, so that $x_2 - x_1 = n\,\Delta x$, the total penetrability can be written in terms of the individual penetrabilities as $P = P_1 P_2 \cdots P_n \approx e^{\gamma_1 + \gamma_2 + \cdots \gamma_n}$. As $n \to \infty$ and $\Delta x \to dx \to 0$, expression (2-107) is obtained.

for the one-dimensional equation. Hence, in this case also, $P \approx e^{-\gamma}$ with

$$\gamma = \frac{2}{\hbar} \int_{r_1}^{r_2} \{2m_0[V(r) - E]\}^{\frac{1}{2}} \, dr \qquad (2\text{-}108)$$

This expression will be useful in the discussion of alpha decay. If the potential is due to the mutual influence of two particles, m_0 represents the reduced mass [see Eq. (2-62)].

    Figure 2-7 illustrates the shape of the wave function which would be obtained if the constants a, b determined by the boundary conditions were substituted into Eqs. (2-96), (2-97), and (2-100). Note that in our example the energy of the particle is not changed and hence the wavelength of the wave function is identical on both sides of the barrier.

FIGURE 2-7 Illustration of the wave function for the barrier shown in Fig. 2-5. The source of particles is located at $x = -\infty$.

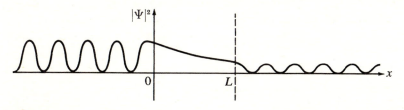

2-2h Parity.   We can see by inspection of Eqs. (2-15) and (2-19) that the substitution $x \to -x, y \to -y$ and $z \to -z$ (abbreviated by $\mathbf{r} \to -\mathbf{r}$ below) will not alter the solutions of Schrödinger's equation if

$$V(-x,-y,-z) = V(x,y,z) \qquad (2\text{-}109)$$

The substitution $\mathbf{r} \to -\mathbf{r}$ is called the parity operation and a potential which has the property expressed in Eq. (2-109) is said to be conservative under the parity operation, or to "conserve parity." It turns out that practically all physical potentials, including those generated by nuclear forces, possess this property.

    For a potential of the form (2-109) the wave function ψ in Eq. (2-19) must have the property[1]

$$\psi(-\mathbf{r}) = +\psi(\mathbf{r}) \qquad (2\text{-}110)$$

or

$$\psi(-\mathbf{r}) = -\psi(\mathbf{r}) \qquad (2\text{-}111)$$

Further, if any system, however complicated, has a wave function of a given type it can never change over to a wave function of the other type (as long as the interactions in the system remain parity-conserving). The wave function

[1] Schiff, 1955, sec. 23.

(2-110) is said to possess even parity, or, briefly, is *even*; the other wave function is *odd*.

The conservation of parity in nuclear interactions places important restrictions on the dynamic nuclear processes (decays and reactions). Therefore it is important to determine the parity of nuclear states by experimental or theoretical means. The parity of a (standing) wave function can usually be recognized from the quantum numbers, as we now show in a special case.

In the example of the particle in a closed cubical box, the parity of the wave function (2-89) is not a definite quantity [since $\psi = 0$ outside the box, it is easy to see that $\psi(x) \neq \psi(-x)$ for $0 < |x| < L$]. This occurs because the location of the box with respect to the origin (see Fig. 2-3a) causes V *not* to have the property (2-109). If the origin is moved to the center of the box, V will have the property (2-109) and the wave function then has the form

$$\psi = \left(\frac{2}{L}\right)^{\frac{3}{2}} \sin\left(\frac{n_x\pi x'}{L} + \frac{n_x\pi}{2}\right) \sin\left(\frac{n_y\pi y'}{L} + \frac{n_y\pi}{2}\right) \sin\left(\frac{n_z\pi z'}{L} + \frac{n_z\pi}{2}\right) \quad (2\text{-}112)$$

where x', y', z' are the coordinates measured with respect to the center of the box ($x' = x - L/2$, etc.). For any odd value of n_x the first sine function becomes

$$\pm\cos\frac{n_x\pi x'}{L} \quad (2\text{-}113)$$

which has *even* parity. For any even value of n_x, the first sine function becomes

$$\pm\sin\frac{n_x\pi x'}{L} \quad (2\text{-}114)$$

which has *odd* parity. Hence the overall parity of the above wave function is even or odd depending on whether or not $(n_x + n_y + n_z)$ is an odd or even integer.

We can also show[1] that the wave function (2-43), applicable to spherical potentials, has the parity $(-1)^l$, where l is the orbital quantum number which determines the orbital angular momentum $[l(l + 1)]^{\frac{1}{2}}\hbar$ of the system.

Having now completed the discussion of those concepts of quantum mechanics which we need for an understanding of nuclear structure, we can return to the subject of nuclear physics.

2-3 NUCLEAR BINDING ENERGY

Every nucleus has a state of lowest energy, the ground state, and higher energy states, called excited states. Much can be learned about nuclear forces from a consideration of nuclei in their ground state, independently of whether these nuclei happen to be stable or have the possibility to decay radioactively. Systematic overall trends can be found in mass, radius, charge, abundance, etc. On closer examination certain periodicities become apparent also. Nuclear models which have been developed to explain these properties can be divided

---

[1] Schiff, 1955, sec. 14.

roughly into semiclassical (*particle*) models, which allow a general understanding of systematic trends and quantum (*wave*) mechanical models which alone give insight into the periodicities. The *liquid-drop model* and the *shell model* are the outstanding representatives of each class and will be described below.

2-3a Definitions.   One of the most important quantities to be considered is the nuclear mass. It is usually expressed in mass units, abbreviated by u, so defined that the mass of one atom of C^{12} is equal to exactly 12.00 ... u.[1] Masses of stable nuclides are listed in App. C.

The difference between the actual nuclear mass and the mass of all the individual nucleons is called the *total binding energy* $B_{tot}(A,Z)$. It represents the work necessary to dissociate the nucleus into separate nucleons or, conversely, the energy which would be released if the separated nucleons were assembled into a nucleus. For convenience, the masses of atoms rather than the masses of nuclei are used in all calculations. This causes no difficulty, except that the binding energy of the atomic electrons should also be considered.[2] For simplicity, though, we will usually omit it. We can therefore write

$$B_{tot}(A,Z) = [ZM_H + NM_n - M(A,Z)]c^2 \qquad (2\text{-}115)$$

where the definitions of the quantities are identical to those of Eqs. (1-1) and (1-2). The average binding energy per nucleon is given by

$$B_{ave}(A,Z) = \frac{B_{tot}(A,Z)}{A} \qquad (2\text{-}116)$$

The following quantities are sometimes convenient, although we will not use them (except in App. C)

$$\text{Mass excess} = M - A \qquad (2\text{-}117)$$

$$\text{Packing fraction} = \frac{M - A}{A} \qquad (2\text{-}118)$$

The work necessary to separate a proton, neutron, deuteron, or alpha particle from a nucleus is called the *separation energy S*. Conversely, this energy is released when such a particle is captured by a nucleus. For a neutron

$$S_n = [M(A - 1, Z) + M_n - M(A,Z)]c^2 \qquad (2\text{-}119)$$

All separation energies can be expressed in terms of the total binding energies of the nuclei involved by substituting the expression for the mass, obtained from Eq. (2-115), into expressions similar to (2-119). We then find, for example,

$$S_n = B_{tot}(A,Z) - B_{tot}(A - 1, Z) \qquad (2\text{-}120)$$

$$S_\alpha = B_{tot}(A,Z) - B_{tot}(A - 4, Z - 2) - B_{tot}(4,2) \qquad (2\text{-}121)$$

---

[1] Before 1960 it was common to set the mass of one atom of O^{16} equal to exactly 16.00 ... atomic mass units (amu). This *physical scale* of atomic masses is not to be confused with the *chemical scale* in which the average mass of one atom of oxygen in the natural isotopic mixture is set equal to 16.00 ....

[2] Evans, 1955, chap. 3, sec. 2. See also Prob. 2-15.

2-3b Average binding energy per nucleon. Saturation and short range of nuclear forces. Experimentally, B_{tot} can be determined from an accurate measurement of M by mass spectrometry or from a determination of S by nuclear reaction studies. The overall trends for B_{ave} are summarized in Fig. 2-8.

FIGURE 2-8 Average binding energy per nucleon versus mass number for the naturally occurring nuclides (and Be^8). Note the scale change on the abscissa at $A = 30$. (By permission from Evans, 1955.)

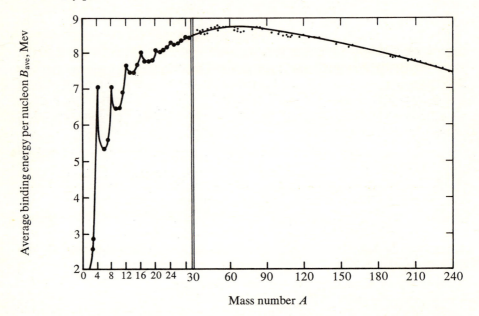

Mass number A

The most striking feature of B_{ave} is its approximate *independence* of A, except for the lightest nuclei. Suppose that the binding energy (bonding energy in chemical terms) of every nucleon to every other nucleon in the nucleus is roughly equal to a constant C. In a nucleus with A nucleons there would then be $\frac{1}{2}A(A-1)$ bonds and hence

$$B_{tot} \approx \tfrac{1}{2}CA(A-1) \tag{2-122}$$

so that

$$B_{ave} \approx \tfrac{1}{2}C(A-1) \tag{2-123}$$

in complete contradiction to Fig. 2-8. The approximate constancy of B_{ave} indicates that each nucleon is not bonded equally to every other nucleon, but rather that nuclear forces between nucleons do not extend to more than a few nucleons. Either the forces must have a very short range of the order of the "diameter" of one nucleon, or they saturate, like chemical bonds. *Saturation* means that the binding, or bonding, energy between one nucleon and the rest of the nucleus reaches a limit once a certain total number of nucleons has been assembled. From Fig. 2-8 it appears that with four nucleons, or more, saturation has set in.

We can find out which of the aforementioned effects is of importance by the following argument. The range of nuclear forces can be inferred from a study of the scattering of two nucleons (p,p or n,p) and from the binding energy of the deuteron.[1] We find that the range is of the order of 2 F, which is comparable to the diameter of a nucleon. This in itself might lead to a constant B_{ave}, if each nucleon were bonded only to its nearest neighbors. But the volume of a nucleus would not vary proportional to A, that is $R \neq R_0 A^{\frac{1}{3}}$ in contradiction to Eq. (1-5). The reason for this is that the nucleons in a given nucleus arrange themselves in such a way as to produce a system of minimum total energy. With the above attractive nuclear force, the lowest *potential* energy is reached if all nucleons crowd into a region so that each one is within about 2 F of the others. The lowest *kinetic* energy is obtained if each nucleon moves in the largest possible nuclear volume.[2] Since the potential energy turns out to be dominant,[3] the nucleus would collapse[4] to a radius of the order of 2 F. Evidently some other effect besides a short force range must occur.

Recent theories of nuclear structure trace saturation to two effects. First, it has been established experimentally that at distances of the order of $\frac{1}{2}$ F the force between nucleons becomes strongly repulsive. We can say that nucleons have a *hard core*. Although this alone would give an $A^{\frac{1}{3}}$ dependence for the nuclear radius, the calculated constant R_0 in Eq. (1-5) comes out too small. Second, the Pauli exclusion principle, which forbids two nucleons of the same kind, e.g., two protons, to occupy states with identical quantum numbers, produces effects which keep nucleons apart from each other.[5]

In summary, a rough consideration of the nuclear binding energy and of the nuclear volume already provides important clues about the nuclear force. Before proceeding to more details, it is instructive to mention another physical system in which the average binding energy per particle is a constant, namely a solid or a liquid. The heat of vaporization Q is the total work necessary to dissociate m grams of the substance into n separated molecules, at a constant temperature. If M_0 is the mass of one molecule

$$m = nM_0 \qquad\qquad (2\text{-}124)$$

[1] See Appendix A.
[2] Compare Eq. (2-90) for one particle. Note that the kinetic energy decreases as L increases.
[3] For a detailed derivation see Blatt and Weisskopf, 1952, pp. 121ff.
[4] Nucleons cannot be regarded as hard spheres with definite diameters. Rather we must consider them as force-exerting entities, which can overlap in accordance with the concepts of quantum mechanics.
[5] For a pictorial and detailed presentation of these points, see Gomes, Walecka, and Weisskopf, 1958.

The average binding energy per molecule is equal to

$$\frac{Q}{n} = \frac{QM_0}{m} \tag{2-125}$$

Experimentally it is found that $Q \sim m$, and Q/m is called the *latent heat of vaporization*. For water at 100°C

$$\frac{Q}{m} = 540 \text{ cal/g} = 2.26 \times 10^{10} \text{ ergs/g}$$

$$M_0 = \frac{18}{6.02 \times 10^{23}} = 2.99 \times 10^{-23} \text{ g}$$

Hence

$$\frac{Q}{n} = 6.75 \times 10^{-13} \text{ ergs} = 0.42 \text{ ev}$$

Comparing this with B_{ave} we see again that atomic and nuclear energies are of the order of ev and Mev, respectively, as indicated at the end of Sec. 2-2f [Eqs. (2-94) and (2-95)] in another connection.

Figure 2-8 shows that in the case of the lightest nuclei, those with a nucleon constitution equal to an integral number of alpha particles have particularly high binding energies per nucleon. This can be understood only on the basis of a quantum mechanical model of nuclear structure in which the dependence of the nuclear force on the intrinsic spin of nucleons is considered. We can appreciate, though, that it is tempting to propose an alpha-particle model for these nuclei, in which the alpha particles are coherent entities and the bonding occurs between them rather than between individual nucleons. Such a model has met with limited success.[1]

The other feature of Fig. 2-8 which we should note is the decrease of B_{ave} toward higher A. This is caused by the increasing influence of the coulomb force, as we will see below.

2-3c Separation energy systematics.   Typical regularities in the neutron separation energies S_n are evident from Fig. 2-9. For a given Z, S_n is larger for nuclei with even N than with odd N. Similarly, for a given N, S_p is larger for even Z than odd Z. The effect is caused by a property of the nuclear force producing extra binding between pairs of identical nucleons in the same state, which have opposite directed (total) angular momenta. This is also the cause of the exceptional stability of the alpha particle structure, mentioned above. In later sections, more evidence for such *pairing* will be given. The difference

$$S_n(A, Z, \text{ even } N) - S_n(A - 1, Z, N - 1) \tag{2-126}$$

[1] Blatt and Weisskopf, 1952, pp. 292ff.

is called the neutron pairing energy and varies from approximately 4 to 2 Mev
with increasing A. Similar values are obtained for protons.

Pairing causes *even-even* nuclei (Z even, N even) to be more stable than
even-odd or odd-even nuclei, and these in turn to be more tightly bound than
odd-odd nuclei. This is also apparent from the abundance systematics of stable
nuclides.

FIGURE 2-9   Neutron separation energies of lead isotopes as a function of neutron
number. (Data from H. T. Tu, "Chart of Mass Differences," *Nuclear Data Sheets*,
vol. 5, set 3, 1963.)

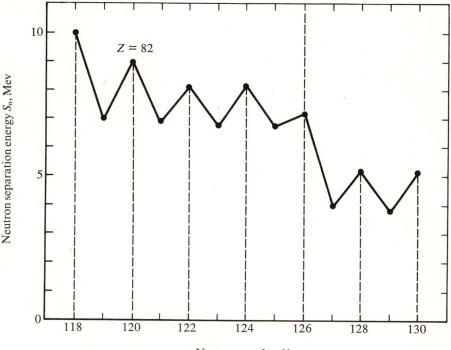

2-3d Abundance systematics of stable nuclides.   The nuclides found on the earth
are either stable or are radioactive with half-lives longer than approximately 10^9
years,[1] since they were produced at least 5×10^9 years ago, according to current
theories. Figure 2-10 presents an N, Z plot for the known stable nuclides,[2]
divided into odd and even isobars. For light nuclides, the average *line of
stability* clusters around $N = Z$; for heavier ones, it deviates from this because
of the increasing importance of the coulomb force. For odd A, only one stable

[1] Decay products of long-lived nuclides are also found. See Sec. 4.2.
[2] See Appendix C.

isobar exists (exceptions $A = 113, 123$). For even A, only even-even nuclides exist (exceptions $A = 2, 6, 10, 14$). A summary of the frequency of occurrence is given in Table 2-2.

FIGURE 2-10   Neutron number versus proton number for stable nuclides. Odd isobars are plotted on the left side and even isobars on the right side. Arrows point along the "magic number" values of N and Z: 20, 28, 50, 82, 126. The odd-odd isobars with $A = 2, 6, 10,$ and 14 are also shown.

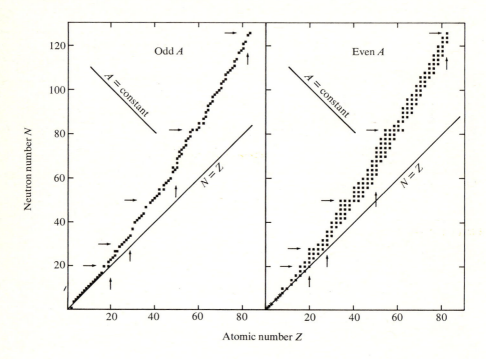

Atomic number Z

Even-even nuclides occur most frequently. If stable nuclei were formed by a process in which increased binding energy produced increased abundance, we could deduce that even-even nuclei are the most stable type of nuclei, i.e., we could equate abundance with stability. This would agree with the conclusions drawn from separation energy systematics. The process of element formation probably was complex, but in one possible formation process, supernova explosions, the binding energy of nuclei does play a dominant role in governing abundance. The belief now is that most nuclides (although not the most abundant) were indeed formed in this process.[1]

[1] For more details of the process of element formation see Smith, 1965, chap. 22, and references given there.

TABLE 2-2 Frequency of occurrence of stable nuclides

N	even	odd	even	odd
Z	even	even	odd	odd
Number of nuclides	160	53	49	4

The relative abundance of isotopes and the cosmic abundance of nuclides also possess interesting regularities. As an illustration, Fig. 2-11 gives the relative isotopic abundance of the element tin ($Z = 50$). The lower relative abundance of the isotopes with odd N is quite apparent. It is again connected with the fact that the process of nuclide formation favored nuclides with higher binding energy. Detailed studies of cosmic abundance lead to the same conclusions.

Particularly high stability and high abundance with respect to neighboring nuclides is associated with nuclides for which N or Z is equal to 2, 8, 20, 28, 50, 82, and 126. Some influence of these *magic numbers* can be noticed by close inspection of Fig. 2-10; other evidence for their existence will be presented later. The magic numbers reflect effects in nuclei very similar to the closing of electronic shells in atoms. There are good reasons why the numbers do not all agree with

FIGURE 2-11 Relative abundances of the tin isotopes as a function of neutron number. The isotopes with $N = 63, 71, 73$ are not stable.

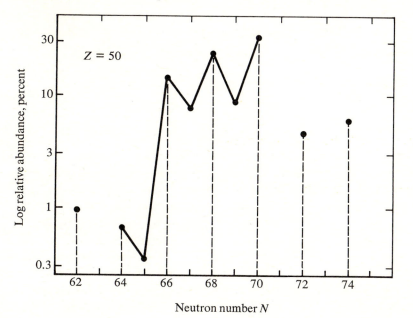

the periods of the periodic table 2, 8, 18, 32, ... Before discussing this *shell* model of nuclei, the liquid-drop model will be reviewed because it is easy to understand and it explains most of the experimental data mentioned so far.

2-4   LIQUID-DROP MODEL. SEMIEMPIRICAL MASS FORMULA

A detailed theory of nuclear binding, based on highly sophisticated mathematical techniques and physical concepts, has been developed by Brueckner and co-workers (1954–1961). A much cruder model exists in which the finer features of nuclear forces are ignored, but the strong internucleon attraction is stressed. It was derived by von Weizsäcker (1935) on the basis of the liquid-drop analogy for nuclear matter, suggested by Bohr. The essential assumptions are (see Sec. 2-3a):

1   The nucleus consists of incompressible matter so that $R \sim A^{\frac{1}{3}}$.

2   The nuclear force is identical for every nucleon and in particular does not depend on whether it is a neutron or a proton.

3   The nuclear force saturates.

Coulomb and quantum mechanical effects are considered separately. From assumptions *2* and *3*, in an "infinitely" large nucleus of A nucleons the main binding energy is proportional to A. Actual nuclei are finite—usually a spherical shape is assumed as in Fig. 2-12—hence nucleons on the surface are not attracted

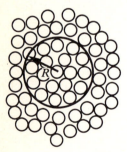

FIGURE 2-12   A spherical nucleus in infinite nuclear matter.

as much as just estimated. A term proportional to the number of nucleons in the surface or proportional to the surface area must be subtracted from the infinite nucleus estimate. Also, the binding will be decreased because the coulomb repulsion acts between all pairs of protons. (Coulomb forces are long range and do not saturate.) In addition, a term must be introduced which tends to give largest binding to nuclei with $N = Z$. This term is a direct consequence of the quantum mechanical behavior of neutrons and protons. Finally, correction terms must be added which give largest binding to even-even nuclei and least binding to odd-odd nuclei and which reflect the shell effects discussed above.

The importance of the model lies in the fact that it explains the empirical features of nuclear mass data. This tells us that the main binding energy term, proportional to A, must be correct. Since this term depends among other things on the assumption of *charge independence* of nuclear forces, we can conclude that n-n, p-p, and p-n nuclear interactions are identical. This important clue about nuclear forces will be supported further.

Keeping in mind expression (2-115), we can write the total binding energy of a nucleus

$$B_{tot}(A,Z) = a_v A - a_s A^{\frac{2}{3}} - a_c \frac{Z(Z-1)}{A^{\frac{1}{3}}} - a_a \frac{(N-Z)^2}{A} \pm \delta + \eta \quad (2\text{-}127)$$

where $\quad a_v A$ = volume term

$-a_s A^{\frac{2}{3}}$ = surface term $\sim$ surface area $4\pi R^2$

$\pm \delta$ = pairing energy term,[1] chosen to be zero for odd-A nuclides; for even-even nuclides the $+$ sign is used, for odd-odd nuclides the $-$ sign applies

η = shell term, positive if N or Z approaches a magic number

The other two terms, coulomb and asymmetry energy, are discussed below.

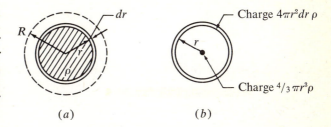

FIGURE 2-13 Coulomb energy of a uniformly charged sphere. (*a*) Actual charge distribution; a layer of thickness dr is added to a sphere of radius r. (*b*) Equivalent charge distribution for purpose of potential energy calculation. The density of charge is called ρ.

$-dr$

R

r

0

Charge $4\pi r^2 dr\,\rho$

r

Charge $\frac{4}{3}\pi r^3 \rho$

(*a*)   (*b*)

2-4a Coulomb energy of a spherical nucleus. Although coulomb forces act between pairs of protons, it is sufficient for the present purpose to consider the nucleus as a uniformly charged sphere of charge Ze and charge density

$$\rho = \frac{Ze}{\frac{4}{3}\pi R^3} \quad (2\text{-}128)$$

We can compute the coulomb energy in the following way. Assume a charged sphere of radius r has been built up, as shown in Fig. 2-13a. The additional work required to add a layer of thickness dr to the sphere can be calculated by assuming the charge $\frac{4}{3}\pi r^3 \rho$ of the original sphere is concentrated at the center

[1] The magnitude of δ is approximately equal to one-half of expression (2-126) as can be seen by substituting Eq. (2-120) for each term and then applying Eq. (2-127).

of the shell (see Fig. 2-13b). The electrical potential energy of the nucleus is therefore

$$V_{\text{coulomb}} = \int_0^R \tfrac{4}{3}\pi r^3 \rho \cdot 4\pi r^2 \, dr\rho \cdot \frac{1}{r}$$

$$= \tfrac{16}{15}\pi^2 \rho^2 R^5$$

$$= \frac{3}{5}\frac{Z^2 e^2}{R} \tag{2-129}$$

using Eq. (2-128). Since, in agreement with a wave function picture, we have assumed the charge of each proton to be "smeared" over the entire nucleus, expression (2-129) contains a spurious "self-energy" term $3e^2/(5R)$ for each proton (found by setting $Z = 1$). Subtracting this term for Z protons gives the correct interaction energy between all pairs of protons

$$V_{\text{coulomb}} = \frac{3}{5}\frac{Z(Z-1)e^2}{R} \tag{2-130}$$

Comparison with Eqs. (1-5) and (1-6) allows us to evaluate the constant a_c in Eq. (2-127)

$$a_c = \frac{3}{5}\frac{e^2}{R_0}$$

$$= 0.62 \text{ or } 0.72 \text{ Mev} \qquad \text{for } R_0 = 1.4 \text{ or } 1.2 \text{ F} \tag{2-131}$$

The coulomb term in Eq. (2-127) occurs with a negative sign because the positive coulomb energy decreases the nuclear binding energy.

2-4b Asymmetry energy.   A very simple model suffices to demonstrate the form of the asymmetry term in Eq. (2-127). Since neutrons and protons obey the laws of quantum mechanics, they must be in definite energy states, similar to those of a closed box (Sec. 2-2f). For ease of calculation, assume that the levels are equidistant with spacing Δ and that as a result of the Pauli exclusion principle there is only one identical nucleon per level. Under the assumption that forces between neutrons are identical to forces between protons except for coulomb effects (see Sec. 2-4), the energy states of neutrons and of protons are expected to be identical.

The *asymmetry energy* is the difference in nuclear energy of a nucleus with neutron and protons numbers N and Z and that of the isobar with neutron and proton numbers both equal to $A/2$. If, to make the former nucleus from the latter, ν protons have to be transformed into neutrons, i.e.,

$$N = \tfrac{1}{2}A + \nu \qquad Z = \tfrac{1}{2}A - \nu \qquad \text{or} \qquad \nu = \tfrac{1}{2}(N - Z)$$

an amount of work equal to

$$\nu^2 \Delta = \tfrac{1}{4}(N - Z)^2 \Delta \tag{2-132}$$

will have to be expended. This can be seen from Fig. 2-14. Note that each of the ν protons will have to be raised in energy by an amount $\nu\Delta$. Since expression (2-132) is always positive, the binding energy of a nucleus will always be less for a nucleus with $N \neq Z$ compared to one with $N = Z$. We can also show that $\Delta \sim 1/A$ by computing the energy E_{max} to which the levels of the nucleus have to be filled to accommodate N neutrons and then setting $\Delta \approx E_{max}/N$.

FIGURE 2-14 Model for the asymmetry term. Neutrons and protons are assumed to have equidistant states of spacing Δ. Crosses represent originally occupied states. In the transfer of three protons to neutron states an energy $3 \times 3\Delta$ will have to be expended.

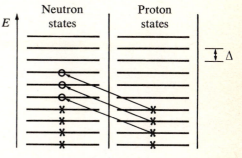

2-4c Mass parabolas. Stability line. With a little rearrangement of Eq. (2-127), we can write the mass of a nucleus [see Eq. (2-115)] in the following way

$$M(A,Z)c^2 = xA + yZ + zZ^2 \mp \delta - \eta \qquad (2\text{-}133)$$

where

$$x = M_n c^2 - a_v + a_a + \frac{a_s}{A^{\frac{1}{3}}}$$

$$y = -4a_a - (M_n - M_H)c^2 \approx -4a_a \approx -94.362 \ MeV$$

$$z = \frac{4a_a}{A} + \frac{a_c}{A^{\frac{1}{3}}}$$

For $A = $ constant, Eq. (2-133) is the equation of a parabola. The minimum mass occurs for $Z = Z_A$ (usually not an integer). The plot of Z_A versus A or N gives the line of greatest nuclear stability. Setting $\partial(Mc^2)/\partial Z = 0$ yields

$$Z_A = \frac{-y}{2z}$$

$$\approx \frac{A/2}{1 + \frac{1}{4}(a_c/a_a)A^{\frac{2}{3}}} \qquad (2\text{-}134)$$

This follows exactly the shape of the empirical *stability line* in Fig. 2-10. By fitting the data we find $\frac{1}{4}(a_c/a_a) = 0.0078$, so that with Eq. (2-131) the expected value of a_a is

$$a_a \approx 20 \text{ to } 23 \text{ Mev} \qquad (2\text{-}135)$$

From expression (2-134), we can recognize that the deviation of the stability line from $N = Z$ or $Z = A/2$ is caused by the competition between the coulomb energy, which favors $Z_A < A/2$, and the asymmetry energy which favors $Z_A = A/2$.

For odd-A isobars, $\delta = 0$, and Eq. (2-133) gives a single parabola, which is shown in Fig. 2-15a for a typical case. We will see later (Sec. 4-6b) that if

$$M(A,Z) > M(A, Z + 1) \qquad \text{beta (electron) decay takes place from } Z \text{ to } Z + 1$$

$$M(A,Z) > M(A, Z - 1) \qquad \text{electron capture and perhaps positron decay[1] takes place from } Z \text{ to } Z - 1$$

$$(2\text{-}136)$$

FIGURE 2-15 Mass parabola for isobars. (a) Odd A nuclei. (b) Even A nuclei. Full circles represent stable nuclides and open circles radioactive nuclides. Along the ordinate, one division is approximately equal to 1 Mev. (By permission from Evans, 1955.)

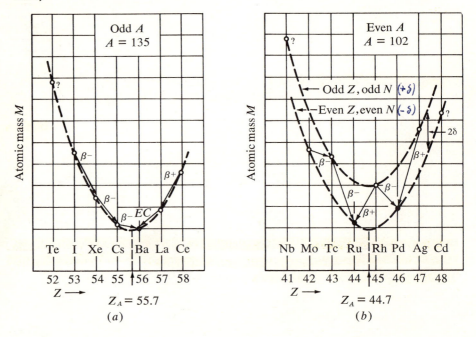

It is clear from Fig. 2-15a that for odd-A nuclides there can be only *one* (stable) isobar for which both these conditions do *not* occur. Figure 2-10 shows this is indeed found. Two exceptions, at $A = 113$ and 123, are no doubt due to the fact that in each case one of the isobars has an exceptionally long half-life (10^{12} years is an experimental lower limit) because the mass differences happen to be exceptionally small.

[1] Positron decay can take place only if $M(A,Z) > M(A, Z - 1) + 2m_0$, where m_0 is the rest mass of an electron. See Eq. (4-122).

For even-A isobars, two parabolas are generated by Eq. (2-133), differing in mass by 2δ. A typical case is given in Fig. 2-15b. Depending on the curvature of the parabolas and the separation 2δ, there can be several stable even-even isobars. Three is the largest number found in nature (see Fig. 2-10). There should be no stable odd-odd nuclides. The exceptional cases H^2, Li^6, B^{10}, and N^{14} are caused by rapid variations of the nuclear binding energy (Fig. 2-8) for very light nuclides, because of nuclear structure effects which are not included in the liquid-drop model. Figure 2-15b shows that for certain odd-odd nuclides both conditions (2-136) are met so that electron and positron decay from the identical nuclide are possible and do indeed occur (see Fig. 4-28, Cu^{64}).

FIGURE 2-16   Summary of liquid-drop model treatment of average binding energy. (By permission from Evans, 1955.)

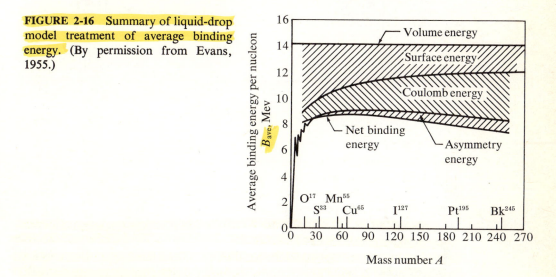

2-4d Summary. Shell effects.

The constants of the semiempirical mass formula (2-127) can be determined by comparison with available data. The "fit" is never perfect, and hence several sets of coefficients have been used. Two such sets are (in Mev; 1 u = 931 Mev)

$$a_v = 14 \qquad a_s = 13 \qquad a_c = 0.60 \qquad a_a = 19 \qquad \delta = 34/A^{\frac{3}{4}}$$
$$a_v = 16 \qquad a_s = 18 \qquad a_c = 0.72 \qquad a_a = 23.5 \qquad \delta = 11/A^{\frac{1}{2}} \qquad (2\text{-}137)$$

The pairing term δ should be roughly equal to one-half of the pairing energy (2-126); the expressions given have some theoretical justification. The contribution of the various terms to B_{ave} is shown in Fig. 2-16.

If we use Eq. (2-127) without the shell model term η to predict neutron separation energies [Eq. (2-120)], we discover interesting regularities in a comparison with experimental data. Figure 2-17 shows the quantity

$$\Delta S_n = S_n(A,Z)_{\text{exp}} - S_n(A,Z)_{\text{calc}} \approx \eta(Z,A) - \eta(Z, A-1) \qquad (2\text{-}138)$$

The increasing binding on approach of the magic numbers (28), 50, 82, and 126 is striking.

FIGURE 2-17 Comparison of observed and calculated neutron separation energies. The effect of shell closures on the neutron separation energy is apparent. (By permission from Evans, 1955.)

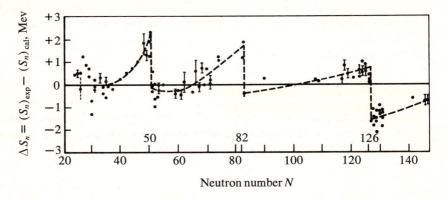

Although we have applied the liquid-drop model only to ground states of nuclei, it can also be used for excited states. These would be produced by oscillations of the nuclear "drop" or by ripples traveling over its surface. This idea has been particularly successful in explaining certain features of nuclear fission, which we will mention in Sec. 5.7.

The liquid-drop model stresses cooperative effects between many nucleons in the nucleus and is the forerunner of the *collective* models of nuclear structure. Implicit in it is a rapid sharing of energy between nucleons, which forms the basis of Bohr's theory of the compound nucleus formation in nuclear reactions.

2-5 SHELL MODEL

The periodic table of the elements is based on regularities in the chemical and physical properties of atoms (valence, types of optical spectra, ionization potential, etc.). The periodicity results from the regular filling of the electronic levels in order of increasing energy, subject to the Pauli exclusion principle which restricts the number of electrons in each sublevel to two. If interelectronic forces are represented approximately by an effective central field, each sublevel is specified by three quantum numbers: the *total* or principal quantum number n_{tot}, the *orbital* or azimuthal quantum number l, and the *magnetic* quantum number m. The latter two have been mentioned in Eqs. (2-44) and (2-45). The total quantum number is given by

$$n_{tot} = n + l \qquad (2\text{-}139)$$

where the *radial quantum number n* is equal to the number of zeros (including the one at $r = 0$) of the radial function $u(r)$ defined in Eq. (2-48). If a sublevel is filled with two electrons, their intrinsic spin directions must have opposite sense in order not to violate the Pauli exclusion principle.

Shortly after the discovery of the neutron, it was proposed that periodicities should also exist in nuclear properties (Bartlett, Guggenheimer, Elsasser, and others, approx. 1933). Regularities were found in abundances and in alpha-decay energies, which implied regularities in nuclear binding energies. It was noted that nucleon numbers 2, 8, and 20 were associated with particular stability. Since these numbers corresponded exactly to the first few periodicity numbers for atomic electrons, a shell structure of nuclei seemed to be indicated. On the whole, though, not much experimental evidence was available to the early shell-model advocates. Also, starting around 1935, successful applications of the liquid-drop model of nuclei and of the compound nucleus model of nuclear reactions suggested that the interaction between nucleons in a nucleus should be so strong as to inhibit any noticeable shell structure.

We can appreciate the preceding statement if we use the form of Heisenberg's uncertainty principle which says that in any experiment of duration t the energy of any system can never be determined more accurately than within an uncertainty Γ where[1]

$$\Gamma t \approx h \qquad (2\text{-}140)$$

Suppose nucleons interact strongly in nuclei and that the mean time between collisions is t. If we would try to determine the energy which a nucleon has in between two collisions, Eq. (2-140) predicts that the result will be uncertain by h/t. The longest time t between collisions that would be reasonable for a nucleus of radius R is of the order of the traversal time

$$t \approx \frac{R}{v} \qquad (2\text{-}141)$$

where v is the speed of the nucleon within the nucleus. According to Eq. (2-27)

$$v = \frac{p}{m_0} = \frac{k\hbar}{m_0} \qquad (2\text{-}142)$$

Since the nucleon is confined to a region of linear dimension R, a relation similar to (2-81) should hold

$$k \approx \frac{\pi}{R} \qquad (2\text{-}143)$$

so that (omitting numerical factors)

$$t \approx \frac{m_0 R^2}{h} \qquad (2\text{-}144)$$

---

[1] This form of the uncertainty principle is derived in another context in Sec. 4.3.

Since t is the longest time between collisions, the minimum energy uncertainty of one nucleon will be

$$\Gamma \approx \frac{h^2}{m_0 R^2} \tag{2-145}$$

But this is exactly of the same order of magnitude as the spacing of nuclear energy levels [see expression (2-90) and its evaluation, Eq. (2-95)]. Nuclear energy states must therefore be so "smeared out" in energy that no well-defined shell structure should occur.

The fallacy in the above argument was recognized by Weisskopf (1951). He pointed out that the Pauli exclusion principle severely restricts the possibility of collisions between nucleons so that times between collisions are much longer than estimated by Eq. (2-141). Consequently, level widths are much smaller than implied by Eq. (2-145).

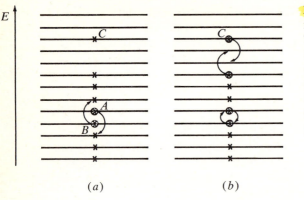

(a)   *(b)*

FIGURE 2-18 Restriction of nucleon-nucleon collisions within a nucleus. (*a*) The indicated collision of particles A and B is not allowed, even though it conserves energy, because the final states are already occupied. (*b*) Possible collisions are those involving exchange of energy levels or excited particles such as C.

We can follow Weisskopf's argument by referring to Fig. 2-18*a*. For simplicity we assume equidistant nuclear levels, each occupied by one nucleon. Consider collisions between particles A and B. In most two-body collisions, the individual kinetic energies are altered even if there is an overall conservation of energy. If, now, the two particles are initially in filled energy states as shown schematically in Fig. 2-18*a*, they cannot collide because the energy states into which they would have to move are already occupied and therefore not available. The only possible collisions are those in which the particles exchange places or in which excited particles, such as C in Fig. 2-18*b*, participate. These are relatively rare. Strong interaction between nucleons is therefore not in contradiction with the existence of shell model effects (i.e., long times between collisions).

2-5a Experimental basis of the shell model.   To date an impressive body of experimental material has been accumulated in which the regularities of nuclear properties are apparent. These point to shell closures at the magic numbers 2,

8, 20, 28, 50, 82, and 126, mentioned in Sec. 2-3d. Before describing the shell model, we present some of the experimental evidence. More details will be given subsequently.

One set of regularities involves nuclear energies directly or, as in abundance data, indirectly (Fig. 2-10). The abundance data are replotted in Fig. 2-19 to bring out the magic numbers more clearly. We can see that the number of isotones (see Sec. 1-2e for definition) is particularly high when N is magic.

FIGURE 2-19 Number of stable isotones as a function of the neutron number. [By permission from B. H. Flowers, *Progr. Nucl. Phys.* **2**: 235 (1952).]

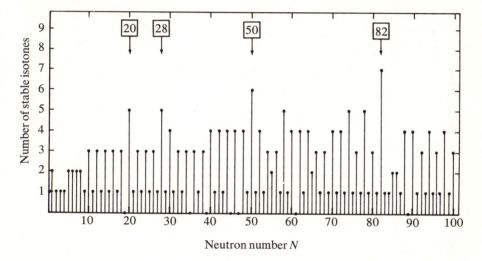

Neutron number N

In connection with Fig. 2-17 we mentioned already that a *high* neutron separation energy is associated with nuclei for which $N = $ (magic number) (see also Fig. 2-9). In addition, Fig. 2-20 shows that for $N = $ (magic number) $+ 1$ the neutron separation energy is particularly *low*. Note that on the abscissa of Fig. 2-20 the neutron number of the final nucleus is plotted. A similar phenomenon is observed in the ionization potential of atoms, which is high for the rare gases and low for the alkalis. Discontinuities are also present in alpha- and beta-decay energies, which reflect discontinuities in nuclear binding energies.[1]

Magic nuclei, being more tightly bound, require more energy to be excited than nonmagic nuclei. This is brought out in Fig. 2-21 in which the excitation energies of the first excited states of even-even nuclei are plotted versus N and Z. The effect of magic numbers extends also to higher excited states; in other words, the spacing between levels is larger for magic nuclei than for other nuclei at comparable excitation energies. Among other things, this results in smaller

[1] This will be discussed further in Sec. 4-5a. See, e.g., Fig. 4-12.

FIGURE 2-20   Separation energy of the last neutron in the nucleus (Z, $N + 1$) for $Z = N$ (even), as a function of the neutron number N of the final nucleus. For $N = 2$, 8, 20, and 28 the separation energy is particularly low. (After M. G. Mayer and J. H. D. Jensen, "Elementary Theory of Nuclear Shell Structure," John Wiley & Sons, Inc., New York, 1955.)

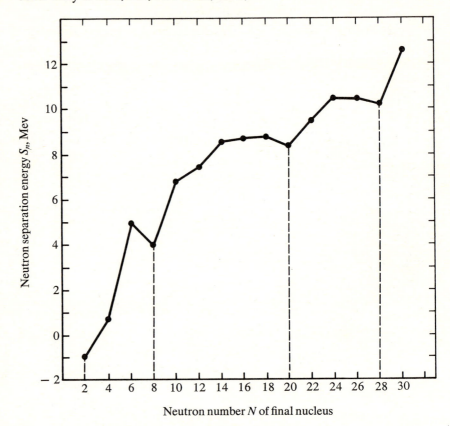

Neutron number N of final nucleus

fast-neutron capture cross sections (Fig. 2-22). A cross section is a quantity proportional to the probability that a nuclear reaction takes place.[1] Crudely speaking, in order for neutron capture to be possible, a neutron of a given kinetic energy must find an empty nuclear level at the correct energy. The further spaced the levels are, therefore, the smaller is the probability of capture.

Regularities are also found in nuclear properties which depend on the total angular momentum and parity of a nucleus, either in the ground state or in an excited state. We will return to this topic at the end of Sec. 2-5c. Finally, we note that the values of nuclear moments which express fine details of the distribution

[1] See Sec. 5-4a.

FIGURE 2-21 Energies of the first excited states of even-even nuclei. [By permission from K. Alder, A. Bohr, T. Huus, B. Mottelson, and A. Winther, *Rev. Mod. Phys.* **28**: 432 (1956).]

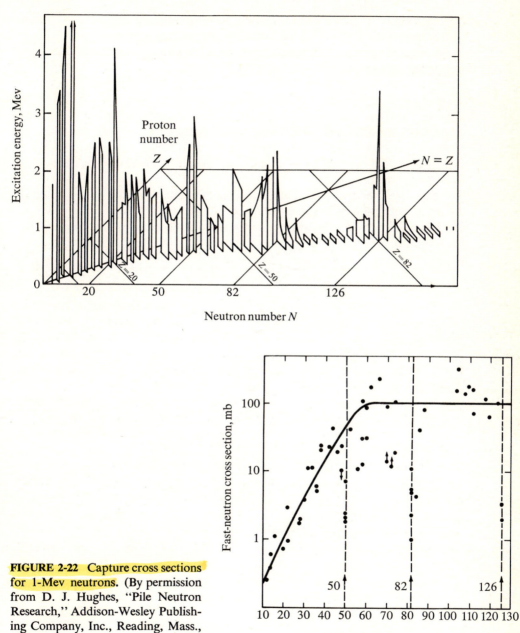

FIGURE 2-22 Capture cross sections for 1-Mev neutrons. (By permission from D. J. Hughes, "Pile Neutron Research," Addison-Wesley Publishing Company, Inc., Reading, Mass., 1953.)

of electric charge and of magnetic dipole strength in a nucleus show regular features. A consideration of these effects is beyond the scope of our treatment, although some mention will be made of them later on.

2-5b Single-particle shell model.   The basic assumption of any shell model is that despite the strong overall attraction between nucleons which provides the binding energy considered in Sec. 2-4, the motion of each nucleon is practically independent of that of any other nucleon. As mentioned in Sec. 2-5, this apparent contradiction is resolved by effects of the Pauli exclusion principle. If all inter-nucleon couplings (called residual interactions) are ignored, we call the model the single-particle shell model. In terms of Schrödinger's equation (2-19), each nucleon is then assumed to move in the same potential. The potential is spherical in the simplest case, but there is good evidence that for nucleon numbers far from closed shells the potential should have an ellipsoidal shape. This condition will be considered later.

For any spherical potential, Schrödinger's equation can be separated as shown in Eq. (2-43), with the angular solutions (2-44) and (2-45). The shape of the potential affects only the radial solution $R(r)$, or more conveniently $u(r) = rR(r)$:

$$-\frac{\hbar^2}{2m_0}\frac{d^2u}{dr^2} + \left[\frac{l(l+1)\hbar^2}{2m_0r^2} + V(r)\right]u = Eu \qquad (2\text{-}146)$$

The boundary conditions on u, in particular as $r \rightarrow \infty$, cause u to be a finite polynomial. The polynomial depends on two quantum numbers, the radial quantum number n and the orbital quantum number l. As mentioned at the beginning of Sec. 2-5, n is equal to the number of nodes of u. Solutions of Eq. (2-146) exist only for definite values of E which also depend on n and l. This situation is quite analogous to the problem of a particle in a closed box. There, too, the boundary conditions determined the quantum numbers of the wave function [Eq. (2-89)] and caused quantization of the energy [Eq. (2-90)].

A common notation for describing the energy states is similar to that used in atomic physics. Whereas in atomic physics each state is specified by the total quantum number n_{tot} [see Eq. (2-139)] and l, in nuclear physics each state is specified by n and l. Also for $l = 0, 1, 2, 3, 4, 5$, we use the spectroscopic letters s, p, d, f, g, h, respectively. A state denoted by $2p$ therefore means that $n = 2$, $l = 1$.

The simplest useful potentials are an infinite square well potential of radius R

$$V = \begin{cases} 0 & r < R \\ \infty & r = R \end{cases} \qquad (2\text{-}147)$$

or a harmonic oscillator potential

$$V = \tfrac{1}{2}m_0\omega^2r^2 \qquad (2\text{-}148)$$

where ω is the frequency of oscillation of the particle of mass m_0. More realistic potentials are a finite square well potential

$$V = \begin{cases} -V_0 & r \le R \\ 0 & r > R \end{cases} \tag{2-149}$$

or a *rounded* well potential which reflects the gradual fall-off of the nucleon density shown in Fig. 1-1.

FIGURE 2-23 Energy levels of nucleons (a) in an infinite spherical square-well potential ($R = 8$ F), (b) in a harmonic oscillator potential. The spectroscopic notation (n, l) and the total occupation number up to any particular level are given. The oscillator number ν, Eq. 2-150, is also shown. (By permission from Burcham, 1963.)

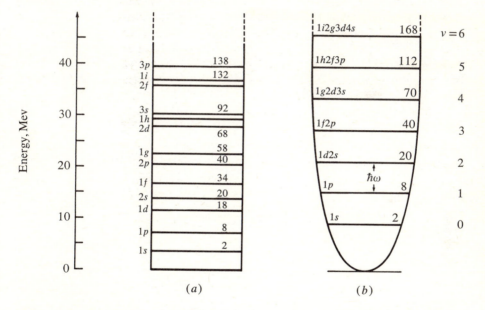

(a)   (b)

The energy levels obtained for the potentials (2-147) and (2-148) are presented in Fig. 2-23a and b, respectively. The spectroscopic notation is given on the left side. Just as in the case of a closed cubical box, the energy of the lowest states does not correspond to zero kinetic energy, and for the same reason. There is no simple mathematical expression for the levels of an infinite square well, but there is one for the harmonic oscillator potential:

$$E = (n_x + n_y + n_z + \tfrac{3}{2})\hbar\omega$$
$$= (\nu + \tfrac{3}{2})\hbar\omega \tag{2-150}$$

where n_x, n_y, n_z are three integral, positive, quantum numbers, which may also have the value zero. The energy levels are equally spaced. As shown in Fig. 2-23b, several levels are degenerate, i.e., the same energy is obtained for more than one set of quantum numbers. The number ν is called the *oscillator quantum number.*

It is also interesting to consider the radial wave function. For the s states ($l = 0$) the mathematical identity of Eqs. (2-53) and (2-22) allows us to write down wave functions for the infinite square well as[1]

$$u = rR(r) = C \sin \frac{n\pi r}{R} \qquad (2\text{-}151)$$

where C is a normalizing constant. For $l \neq 0$ more complicated functions are obtained.[2] For any finite well for which $V(r \rightarrow \infty) \rightarrow 0$, $u(r \rightarrow \infty)$ has the following form for a bound state ($E < 0$)

$$u \sim e^{-\kappa r} \qquad (2\text{-}152)$$

where κ is defined by $\frac{1}{2}\hbar^2\kappa^2/m_0 = |E|$. A few radial wave functions for the finite square well are sketched in Fig. 2-24.

FIGURE 2-24 Schematic radial wave functions for a *finite* square well. For $r \rightarrow 0$, $R(r) \sim r^l$ and for $r \rightarrow \infty$, $R(r) \sim r^{-1}e^{-\kappa r}$, where $\frac{1}{2}\hbar^2\kappa^2/m_0 = |E|$, the energy of the level below $V(r \rightarrow \infty) = 0$. See problem 2-13.

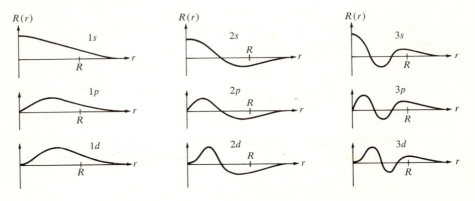

In accordance with the Pauli exclusion principle each level can be filled with identical nucleons such that no two nucleons have the same set of quantum numbers

$$n \qquad l \qquad m \qquad m_s \qquad (2\text{-}153)$$

where $m_s = +\frac{1}{2}$ or $-\frac{1}{2}$ is a quantum number specifying the direction of the

---

[1] See Eqs. (2-80) and (2-81), but note that $u(0) = 0$, because $R(0)$ must be finite independent of the form of $V(r)$.

[2] The first few functions are

$$u(l = 1) = (\sin \rho)/\rho - \cos \rho$$
$$u(l = 2) = [(3/\rho^2) - 1] \sin \rho - (3/\rho) \cos \rho$$

where $\rho = f(n)\pi r/R$, $f(n) =$ numerical factor depending on n. It can also be shown that for $r \rightarrow 0$, $R(r) \sim r^l$. See Schiff, 1955, sec. 15.

intrinsic nucleon spin. For example, if $l = 3$ the possible values of m are -3, $-2, -1, 0, 1, 2, 3$ and in each *magnetic sublevel* two nucleons may be placed with $m_s = +\frac{1}{2}$ and $-\frac{1}{2}$, respectively. The maximum occupation number in this case is therefore $2 \times 7 = 14$; in general it is equal to $2 \times (2l + 1)$. Figure 2-23 gives the total occupation number *up* to any particular level for the two potentials shown. We would expect that whenever an (n,l) level is completely filled that the nucleus should have particularly high stability, because the number of nucleons is even and the maximum pairing effect comes into play. Also, if the gap to the next (unfilled) energy state is large, a larger energy would be required to excite the nucleus than if the gap is small. Therefore, magic-number effects should occur at the major shell gaps. Although the magic numbers 2, 8, and 20 are easily reproduced (see Fig. 2-23), the others (28, 50, 82, 126) are not apparent. Even if the more realistic potential (2-149) or a rounded well potential is used, this difficulty cannot be overcome. Since all the early shell models used these types of potential, they were unable to fit the magic numbers and were thought to be of limited usefulness.

2-5c Spin-orbit coupling model. It is to the credit of Mayer and of Haxel, Jensen, and Suess (1949) that they recognized independently the missing ingredient in the shell model presented so far. They proposed that a strong interaction should exist between the *orbital* angular momentum and the *intrinsic spin* angular momentum of each nucleon. According to the quantum-mechanical coupling rules for angular momenta, the total angular momentum $j\hbar$ formed by the vector addition of the orbital angular momentum $l\hbar$ and the intrinsic spin $s\hbar$ must be such that j is restricted to the values[1]

$$j = l + \tfrac{1}{2} \quad \text{or} \quad j = l - \tfrac{1}{2} \tag{2-154}$$

If a strong spin-orbit interaction exists, a different energy is associated with each of these two values of j, giving rise to a spin-orbit splitting of the levels. In the spectroscopic notation, the j value is placed as a subscript to the (n,l) symbol. For example, in the case of the $1p$ shell, we obtain a splitting between the $1p_{\frac{1}{2}}$ and $1p_{\frac{3}{2}}$ levels. It was found empirically that in nuclei the energy level with the larger j value always lies below that with the smaller j value.

Figure 2-25 shows the effect of spin-orbit splitting on the energy levels of a finite, rounded-well potential. The maximum occupation number for each level $(n,l)_j$ is given by $2j + 1$ since according to the rules of quantum mechanics each vector angular momentum $v\hbar$ can have the projections $m_v\hbar$ along any given axis where

$$m_v = -v, -v + 1, -v + 2, \ldots, v - 2, v - 1, v \tag{2-155}$$

and v is an integer or half-integer.

---

[1] Actually each angular momentum vector of the form $v\hbar$ has a magnitude $[v(v + 1)]^{\frac{1}{2}}\hbar$, but $v\hbar$ is the maximum value of the vector component along any given direction. Therefore, we often call $v\hbar$, or even v, the *angular momentum*.

We can see from Fig. 2-25 that if the magnitude of the spin-orbit splitting is properly adjusted, the major shell breaks occur at the experimentally determined magic numbers. Furthermore, if we assume that the angular momentum of odd-A nuclei is determined solely by the odd-nucleon number, a remarkable agreement is obtained between the ground state spins (and parities) of these nuclei and the predictions of the spin-orbit coupling model. This agreement is

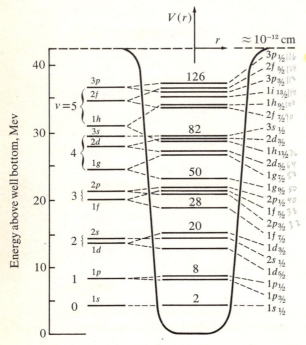

FIGURE 2-25 Energy levels in a rounded potential well including a strong spin-orbit splitting. [By permission from R. J. Blin-Stoyle, *Contemp. Phys.* **1:** 17 (1959).]

shown in Fig. 2-26a and b for odd-N and odd-Z nuclei. Considering also the empirical finding that all even-even nuclei have zero ground-state spins, it follows reasonably that in the ground state of any nucleus the net angular momentum associated with an even N or Z is equal to zero.

A natural consequence of the spin-orbit shell model is that, close to the major shell breaks, levels of large spin difference lie close together. If, for example, near nucleon number 50 a nucleon occupies a $2p_{\frac{1}{2}}$ level, we expect an excited $1g_{\frac{9}{2}}$ state nearby. This gives rise to numerous *isomeric states*[1] near the shell breaks (see Fig. 2-26c). The order of the levels is sometimes reversed by small binding energy effects.

---

[1] These long-lived states are caused by large angular momentum differences and small energy differences with respect to the ground state. This will be discussed further in Sec. 4-4.

2-5d Other nuclear models.   Although the spin-orbit shell model has had one of the most stimulating effects on nuclear structure physics, the simple form given above cannot be sufficient. For example, the model cannot explain why even-even nuclei always have a zero ground-state spin, or more generally, why any even number of identical nucleons couples to zero ground-state spin. Evidently there is a (residual) nucleon-nucleon interaction which favors the

FIGURE 2-26   (*a, b*) Angular momentum quantum number *I* for the ground states of odd-*A* nuclei, plotted versus the odd-nucleon number *N* or *Z*.   (*c*) Number of cases of isomerism in odd-*A* nuclei versus the odd-nucleon number. Dashed vertical lines on the abscissa indicate the magic numbers 2, 8, 20, 28, 50, 82 and 126, predicted by the shell model of Mayer and Haxel, Jensen and Suess. The detailed spectroscopic assignments (see Sec. 2-5c) are also given.

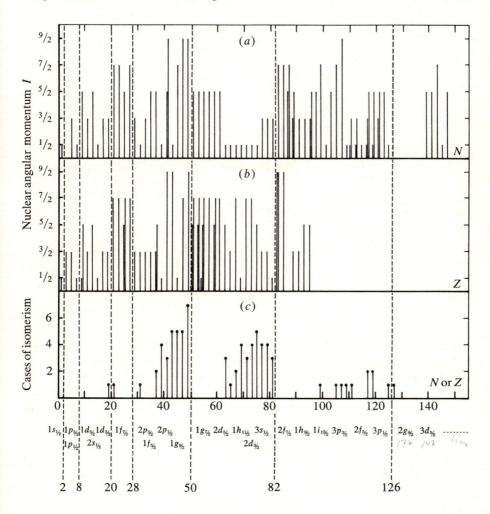

pairing of nucleons with opposing angular momenta. An attractive interaction between nucleons must therefore be added to the single nucleon spin-orbit interaction, which gives rise to the pairing energy already encountered in Sec. 2-3c. From detailed theoretical considerations, it appears likely that the magnitude of the pairing energy increases with the l value of the pair and that for this reason the high-spin states ($h_{\frac{11}{2}}, i_{\frac{13}{2}}$) predicted by the spin-orbit model (Fig. 2-25) are not found in odd-A ground-state spins. For example, the energy of the state consisting of six nucleons filling the $2d_{\frac{5}{2}}$ shell plus one $1h_{\frac{11}{2}}$ nucleon is higher than the energy of five $2d_{\frac{5}{2}}$ nucleons plus two paired $1h_{\frac{11}{2}}$ nucleons, even though the $2d_{\frac{5}{2}}$ level lies below the $1h_{\frac{11}{2}}$ level.

The properties of the pairing interaction can be made plausible by a semiclassical argument. Assume there is an attractive *residual* force between two like nucleons, i.e., a force beyond that already taken into account in the shell-model potential. If the force is of short range and attractive, the total energy of the nucleus will be lowered if the two like nucleons are as close together as possible. In classical language, the two nucleons should collide as often as possible in their orbits. In quantum language, the wave functions should overlap as much as possible. This situation is obtained if (1) the two particles are in the same (n,l) state and (2) their orbital angular momenta $l_1\hbar$ and $l_2\hbar$ are oppositely directed, as shown in Fig. 2-27b. The particles will then collide most often, or more specifically, their wave functions overlap best.[1]

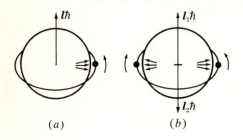

(a)   (b)

FIGURE 2-27 Distorting effect of extra nucleons on a closed shell nucleus. (a) One extra nucleon. (b) Two extra nucleons with opposite orbital angular momenta.

If the pairing forces were not of very short range, but extended perhaps over the entire nucleus, there would be no energetic favoring of close distances between the two interacting nucleons, because their interaction energy would be approximately independent of their separation. In quantum language, the exact overlap of the wave functions would not be so important in that case and there would be no reason why states with $l_1 + l_2 = 0$ should be favored over other orientations of the angular momentum vectors. Since, *experimentally*, states

---

[1] Since the intrinsic spin of the particles has been ignored in this argument, the Pauli exclusion principle forbids the two like particles in the same (n,l) state to have the same m quantum numbers. Therefore the angular momentum vectors are not allowed to be parallel, even though this would give equally good overlap of the wave functions.

with $l_1 + l_2 = 0$ are favored, we can reverse this argument and conclude that the pairing force is of short range compared to the nuclear radius.

Another feature not included in the simple shell model is a distorting effect of the "outermost" nucleons on the other nucleons in a nucleus. Assume that we add a single nucleon to a closed shell nucleus. This nucleon will usually have a high l value (Fig. 2-25) and therefore its wave function will peak close to the nuclear radius (Fig. 2-24). In terms of a classical picture we can say that the nucleon circles around the closed-shell "core" of nucleons as shown in Fig. 2-27a. The strong attraction between the nucleon and the core will distort the core. The core exerts a centripetal force on the nucleon and the reaction to this force, the centrifugal force, acts on the core. If two nucleons are outside the core, they will run in the same orbits (but in opposite sense, because of the pairing effect). Therefore, the distortion of the core will increase. If more and more nucleons are added outside of the closed-shell core, a point will be reached at which the core is permanently deformed, with an accompanying effect on the orbits. Although a detailed consideration is beyond our treatment, one consequence may be noted.

A quantum-mechanical body which has an axis of symmetry such as an ellipsoid of revolution can undergo rotations about an axis perpendicular to the symmetry axis. The energy level spectrum of such a "rotator" is quite character-istic. It is a remarkable fact that there are many cases of even-even nuclei in which this spectrum has been found. This leads us to the conclusion that permanently deformed nuclei indeed exist, as suggested above.

The energy spectrum can be "derived" by a semiclassical argument. The classical kinetic energy of a rotating body is equal to

$$E = \tfrac{1}{2}\mathscr{I}\omega^2 \tag{2-156}$$

where $\mathscr{I}$ is the moment of inertia of the body about the axis of rotation and ω is the angular frequency of rotation. In terms of the angular momentum $L = \mathscr{I}\omega$

$$E = \frac{L^2}{2\mathscr{I}} \tag{2-157}$$

Proceeding from a classical to a quantum mechanical model, we must replace L^2 by $I(I + 1)\hbar^2$ [see Eq. (2-52)] where, for even-even nuclei, I is an even integer with the lowest value zero:

$$E = \frac{I(I + 1)\hbar^2}{2\mathscr{I}} \tag{2-158}$$

In Fig. 2-28 we compare this simple expression with a typical experimental spectrum. The discrepancies at higher excitation energies can be understood. Because the classical rotation frequency increases with increasing I value, the nucleus will be slightly deformed by the larger centrifugal force and its moment of inertia will become larger. According to Eq. (2-158), the excitation energy will decrease, exactly as found (Fig. 2-28).

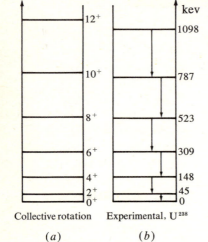

FIGURE 2-28 Energy spectrum of a deformed nucleus. (a) Theoretical spectrum of a quantum-mechanical rotator. (b) The spectrum of U^{238}. [By permission from F. S. Stephens, *Phys. Rev. Letters* **3**: 435 (1959) and Burcham, 1963.]

From the foregoing discussion it appears that the single-particle shell model is valid particularly near the closed shells. As the nucleon numbers deviate more and more from the magic numbers, cooperative effects appear between nucleons. These are most easily incorporated into *collective models*, in which rotational motion, discussed above, and vibrational motion are built in from the start. Theoretical advances have shown, though, that collective effects can also be obtained by modifications of the shell model. It is then necessary to include in the shell model potential the following terms:

A dominant spherical potential
A spin-orbit interaction
An interaction of rather short range, which tends to make the nucleus spherical and tends to pair up nucleons
A long-range term which tends to distort the nucleus

Any theory of nuclear forces must include these features. It is our hope that the meson theory of nuclear forces (Chap. 6) will be able to explain more and more of these properties.

2-6 ENERGY LEVELS OF NUCLEI

In any finite potential well, such as that shown in Figs. 2-25 or 2-29, there are bound energy levels, corresponding to states of energy[1] $E < 0$, and unbound or virtual levels corresponding to states of energy $E > 0$ (see Sec. 2-2c). Virtual

[1] As previously, we assume $V(r \rightarrow \infty) = 0$.

states cannot be understood on any classical model. They result from the fact that the de Broglie wave of a nucleon is reflected at the edge of the potential well even though its total energy exceeds the potential energy everywhere. If the de Broglie wavelength happens to be such that approximately standing waves are formed within the potential well, the amplitude of the wave function can be very large within the well, and a virtual state occurs. This means there is a large probability for a nucleon of quite well-defined energy to be found within the nucleus.

FIGURE 2-29 Schematic representation of nuclear levels. (By permission from Burcham, 1963.)

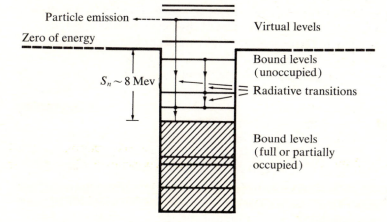

The reflection coefficient of a potential step may be derived using the method illustrated in Sec. 2-2g. In Fig. 2-5, let $L \rightarrow \infty$ and assume $E > V_b$. Then

$$\psi_{\mathrm{I}} = \psi_{\mathrm{I}\rightarrow} + \psi_{\mathrm{I}\leftarrow} = a_{\mathrm{I}}e^{ikx} + b_{\mathrm{I}}e^{-ikx} \qquad (2\text{-}159)$$

$$\psi_{\mathrm{II}} = \psi_{\mathrm{II}\rightarrow} = a_{\mathrm{II}}e^{ik'x} \qquad (2\text{-}160)$$

where $\qquad k' = \left(2m_0 \dfrac{E - V_b}{\hbar^2}\right)^{\frac{1}{2}} \qquad$ and $\qquad k = \left(\dfrac{2m_0 E}{\hbar^2}\right)^{\frac{1}{2}} \qquad (2\text{-}161)$

Applying boundary conditions at $x = 0$ yields

$$a_{\mathrm{I}} + b_{\mathrm{I}} = a_{\mathrm{II}} \qquad \text{from} \qquad \psi_{\mathrm{I}} = \psi_{\mathrm{II}}$$

$$a_{\mathrm{I}} - b_{\mathrm{I}} = a_{\mathrm{II}} \frac{k'}{k} \qquad \text{from} \qquad \frac{d\psi_{\mathrm{I}}}{dx} = \frac{d\psi_{\mathrm{II}}}{dx}$$

From these two equations, we find immediately that the reflection coefficient is equal to

$$\frac{|b_{\mathrm{I}}|^2}{|a_{\mathrm{I}}|^2} = \left(\frac{1 - k'/k}{1 + k'/k}\right)^2 \qquad (2\text{-}162)$$

For $k'/k \ll 1$, this is approximately equal to $1 - 4k'/k$. If $E/V_b = 1.01$, the reflection coefficient is about 0.7.

We will show later on (see Sec. 4-3), in connection with the *dynamic* nuclear properties, that the nucleon leaves the virtual state after some mean life τ, causing it to have a width for particle emission given by[1]

$$\Gamma = \frac{\hbar}{\tau} \tag{2-163}$$

As indicated in Fig. 2-29, a virtual state can also decay to a lower state by gamma emission,[2] analogous to the emission of light by excited atoms. Bound levels can decay only by gamma emission.

According to the single-particle shell model, a given nucleus (Z,N) consists of Z protons and N neutrons normally placed into bound levels (protons and neutrons each into their own potential well) in accordance with the Pauli exclusion principle. In the ground state of the nucleus, all nucleons are in their lowest energy states. The simplest excited states of the nucleus are then formed by promoting the outermost (or least bound) nucleon to a higher state. The corresponding excitation spectrum of the nucleus is called the single-particle level spectrum, with levels as shown in Fig. 2-25.

An actual level spectrum is shown in Fig. 2-30. Even though it is practically certain that not all levels have as yet been found, we can see that there are already many more levels than expected from the single-particle spectrum of Fig. 2-25. Present theories indicate that the single-particle spectrum is actually *dissolved* among many levels each of which consists of complicated excitations of more than one particle. The distribution of the single-particle levels is indicated schematically in Fig. 2-30.

Figure 2-30 also shows that the density of energy states (number of states per unit energy interval) increases with increasing excitation energy. This occurs because a given excitation energy can be reached by a greater variety of excitations of nucleons as the energy increases.[3] Incomplete theoretical arguments are available to predict the energy variation of level density, yielding[4]

$$\rho = 1/\Delta = \rho_0 e^{a\sqrt{E}} \tag{2-164}$$

where ρ = level density
Δ = spacing between levels
$\rho_0 \approx 1$ to 0.01 $(\text{Mev})^{-1}$, for $A = 20$ to 200
$a \approx 1$ to 7 $(\text{Mev})^{-\frac{1}{2}}$, for $A = 20$ to 200

[1] Compare Eq. (2-140). Equation (2-163) will be derived in Sec. 4-3.
[2] For unusual exceptions, see Sec. 4-4e.
[3] The density of single-particle levels also increases with excitation energy, but this plays a relatively minor role in the increase of actual level density.
[4] Burcham, 1963, sec. 15.3.4; Blatt and Weisskopf, 1952, chap. 8, sec. 6.

FIGURE 2-30 Level spectrum of Sc⁴¹.
Note the change in energy scale between
4 and 5 Mev. The angular momentum
l_p of the last odd proton is shown. The
splitting up of the single particle
structure into the actual levels is also
indicated. [Adapted from K. Way,
A. Artna, and N. B. Gove, (eds.),
"Reprint of Nuclear Data Sheets,
1959–1965," Academic Press, Inc., New
York, 1966.]

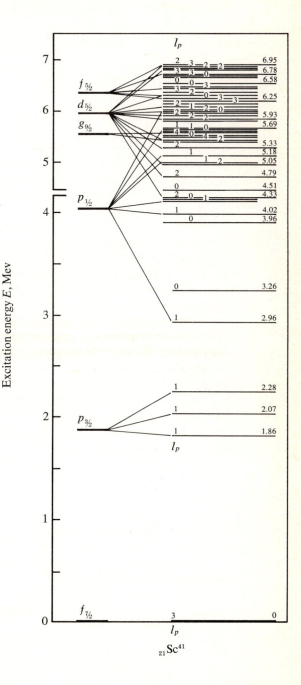

Although, with a very few exceptions, highly excited states of nuclei cannot be described in detail, definite regularities have been observed for low-lying states. For *odd-A nuclei*, low-lying states can be related to the shell model, especially if the nucleon numbers are close to magic numbers. Not only energy values, but also spins and parities can be explained. For nucleon numbers between magic values, it is generally more profitable to describe the states in terms of a collective model. For *even-even nuclei*, excited states can only be produced by breaking up a pair of nucleons. This requires so much (pairing) energy that even the lowest excited states consist of complicated excitations, which usually are more easily described by the concepts of *vibrations and rotations of the nucleus* than in shell model terms. The regular pattern of the first excited states of these nuclei is apparent from Fig. 2-21. Such a simple pattern is not found for odd-*A* nuclei. For *odd-odd nuclei* the lowest excited states can be explained by assuming that the odd protons couple to an angular momentum $I_p\hbar$ and the odd neutrons to $I_n\hbar$. The angular momentum $I\hbar$ of the nucleus is then given by

$$I = I_p + I_n \tag{2-165}$$

where the quantum mechanical addition rules obtain, i.e., $I = |I_p - I_n|, \ldots$ $|I_p + I_n|$ in integral steps. The ground state is usually formed if the intrinsic spin of the last odd neutron is parallel to the intrinsic spin of the last odd proton. The lowest excited states consist of other orientations of I_p and I_n.

A study of the level structure of odd-odd nuclei requires, therefore, that the nuclear force described at the end of Sec. 2-5d must contain yet another term, which tends to align the intrinsic spins of nucleons. The simplest odd-odd nucleus is H^2. It has $I_p = \frac{1}{2}$ and $I_n = \frac{1}{2}$. In its ground state $I = 1$, and in its (virtual) first excited state $I = 0$. The energy difference between these states is 2.3 Mev, which gives the order of magnitude of the spin-spin interaction. We must not confuse the latter with the pairing interaction, which tends to align the orbital angular momenta of two identical nucleons so that the total angular momentum is zero.

2-7 CHARGE SYMMETRY AND CHARGE INDEPENDENCE OF NUCLEAR FORCES

The derivation of the semiempirical mass formula assumed that the main binding energy between nucleons is independent of their nature, i.e., that *p-p*, *n-n*, and *n-p* nuclear forces are identical. Since the semiempirical mass formula is not accurate, it is of interest to present more direct evidence for this *charge independence* of nuclear forces.

Pairs of nuclei for which $Z_1 = n$, $N_1 = n'$, and $Z_2 = n'$, $N_2 = n$ (where n and n' are integers) are called *mirror nuclei*. The most common ones have $n' = n - 1$ or $n' = n - 2$. Figure 2-31 compares the level spectra of such a pair, Li^7 ($Z = 3$, $N = 4$) and Be^7 ($Z = 4$, $N = 3$). Not only energies of the states, but also the angular momenta and parities, agree closely. We can infer from this that nuclear forces are charge symmetric, i.e., *p-p* and *n-n* forces are identical.

Consider the schematic presentation of Li^7 and Be^7 shown in Fig. 2-32. In each nucleus, the nuclear bonds are identical except for those of the (last) odd nucleon. Therefore, if the excitation spectra of these nuclei are identical, we can state

$$3E_{n\text{-}n} + 3E_{p\text{-}n} = 3E_{p\text{-}p} + 3E_{p\text{-}n} \qquad (2\text{-}166)$$

where E represents the bond-energy difference between the excited and the ground

FIGURE 2-31   Comparison of the energy levels of Li^7 and Be^7. Uncertain levels are shown in dashed lines. Corresponding levels are connected by dotted lines. Measured spins and parities are given. [Adapted from T. Lauritsen and F. Ajzenberg-Selove, *Nucl. Phys.* **78**: 1 (1966).]

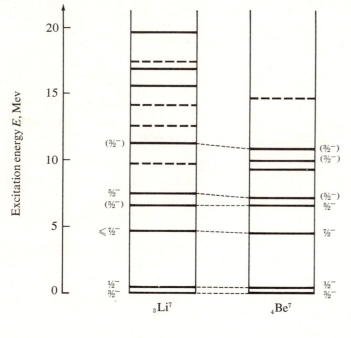

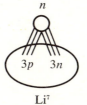

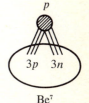

FIGURE 2-32   Comparison of the odd-nucleon bonds in Li^7 and Be^7.

states. Hence

$$E_{p\text{-}p} = E_{n\text{-}n} \tag{2-167}$$

It is obviously tempting to conclude that the bond energies themselves are also equal. That this is so can be shown from the masses of the mirror nuclei in their ground state. If, indeed, nuclear forces are charge symmetric, we can see

FIGURE 2-33 Mass difference between mirror nuclei versus $(Z_1 - 1)A^{\frac{1}{3}}$, according to Eq. (2-168). (Data from C. E. Gleit, C. W. Tang, and C. D. Coryell, "Beta-Decay Transition Probabilities," Nuclear Data Sheets, vol. 5, set 5, 1963.)

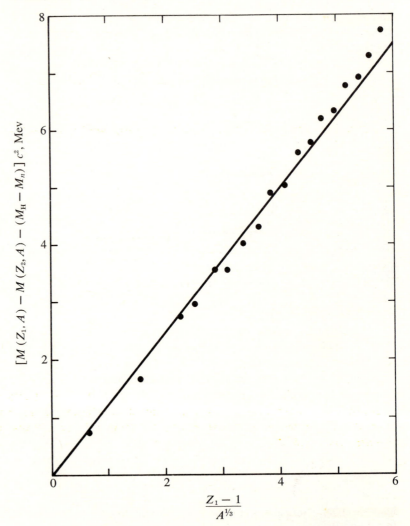

from Fig. 2-32 or from the semiempirical mass formula [Eq. (2-127)] that the mass difference between two mirror nuclei with $Z_1 = Z_2 + 1$ is

$$[M(Z_1,A) - M(Z_2,A)]c^2 = [Z_1(Z_1 - 1) - Z_2(Z_2 - 1)]\frac{3e^2}{5R} + (M_H - M_n)c^2$$

$$= (Z_1 - 1)\frac{6e^2}{5R} + (M_H - M_n)c^2 \qquad (2\text{-}168)$$

The left-hand side of this equation can be determined either directly or from the positron decay energy between the two isobars (see Sec. 4-6b). The right-hand side can be calculated from Eq. (1-5) and agrees with the left side, so that the assumed charge symmetry of nuclear forces is indeed valid. Figure 2-33 summarizes the results of many measurements. A plot of the mass difference, corrected for $(M_H - M_n)c^2 (= -0.78$ Mev), versus $(Z_1 - 1)/A^{\frac{1}{3}}$ should pass through the origin if charge symmetry is valid.[1] The deviation apparent on Fig. 2-33 results from shell-model effects, which give slightly different effective radii to the nuclei (Z_1,A) and (Z_2,A) because the wave functions of the protons are different.

Interesting similarities between levels are also found if we compare masses of *mirror triads*

$$(Z_1 = n - 1, N_1 = n + 1) \qquad (Z_2 = n, N_2 = n) \qquad (Z_3 = n + 1, N_3 = n - 1)$$

in their ground and excited states. Figure 2-34 gives the states of the triad Be^{10}, B^{10}, C^{10} for which $n = 5$, after correcting each mass for the coulomb energy. Again we see that the levels of the mirror nuclei Be^{10} and C^{10} are practically identical,[2] but only certain levels in B^{10} correspond to these states. If we analyze this by considering the bonds involved, as in Fig. 2-35a, we find that for the identical levels in Be^{10}, B^{10}, and C^{10}, p-p, n-n and n-p forces must be identical, but also there are many more levels in B^{10} than in Be^{10} and C^{10}. This is a direct consequence of the Pauli exclusion principle.

To gain some understanding, let us consider the simplest mirror triad, n^2, H^2, He^2. (Although n^2 and He^2 are not stable structures, the argument given is correct.) In the lowest states, which are $1s$ states, the Pauli exclusion principle requires opposing intrinsic spins in n^2 and He^2, but not in H^2. Therefore, the n-p system has more states than the n-n or p-p system. It is also suggestive from this example that charge independence of nuclear forces holds only if the nucleons are in identical states as far as their angular momenta (orbital and intrinsic spin) are concerned. This is confirmed by a detailed analysis of n-p and p-p scattering[3] and is a further important feature of the nuclear force.

[1] The plot also allows a determination of the radius constant R_0 [Eq. (1-5)]. See Prob. 2-26.
[2] Probably, not all low-lying levels of C^{10} have been found.
[3] See Appendix A, Secs. A-3 to A-5.

FIGURE 2-34 Comparison of the mirror triad Be^{10}, B^{10}, and C^{10}. Each mass has been corrected for the coulomb energy, and the neutron-proton mass difference. Uncertain levels are shown in dashed lines. Corresponding levels are connected by dotted lines. Measured spins and parities are given. [Adapted from T. Lauritsen and F. Ajzenberg-Selove, *Nucl. Phys.* **78**: 1 (1966).]

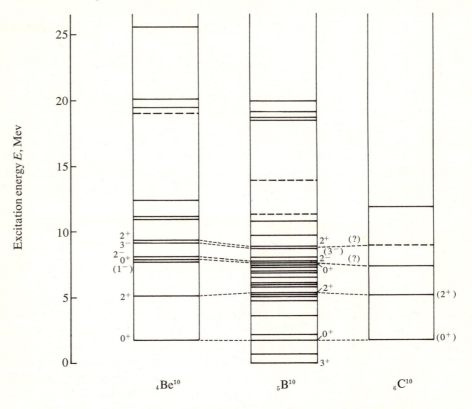

FIGURE 2-35 (*a*) Comparison of bonds in Be^{10}, B^{10}, C^{10}. (*b*) Comparison of states of n^2, H^2, He^2.

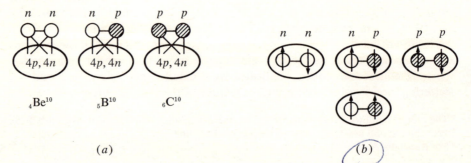

PROBLEMS

2-1 Prove that a particle of rest mass m_0 and kinetic energy T always has more momentum than a photon of energy $h\nu = T$, independent of the value of T.

2-2 It is desired to use the nonrelativistic expression for kinetic energy in order to calculate the speed of a particle of mass m_0. What is the highest allowed energy of the particle (in units of $m_0 c^2$) so that the calculated speed is correct within 1 percent?

2-3 What is the speed of a particle whose kinetic energy is equal to its rest energy?

2-4 Calculate the de Broglie wavelengths of a 10-ev electron, a 10-Mev alpha particle, and a 10-g bullet moving with a speed of 1,000 m/sec.

2-5 At what rate (in g/sec) is matter converted into energy in a nuclear reactor which produces 2 megawatts of power?

2-6 (a) A discus thrower spins a disc weighing 3 kg in a radius of 0.5 m at 1 rps. What is the angular momentum of the disc in units of $\hbar$? (b) If a neutron were to move inside of a nucleus in a 5-F radius with 3 units of orbital angular momentum, what would its speed be? (This classical picture of nuclear motion is incorrect and must be replaced by the wave-function concept.)

2-7 In spherical coordinates the expression $\nabla^2 \psi$ in Eq. (2-19) is

$$\frac{1}{r^2}\frac{\partial}{\partial r}\left(r^2 \frac{\partial \psi}{\partial r}\right) + \frac{1}{r^2 \sin\theta}\frac{\partial}{\partial\theta}\left(\sin\theta \frac{\partial\psi}{\partial\theta}\right) + \frac{1}{r^2 \sin^2\theta}\left(\frac{\partial^2\psi}{\partial\varphi^2}\right)$$

(a) Show that a substitution of the form (2-43) separates the variables. (b) Show that the radial equation can be put into the form (2-47) by appropriate definition of the separation constant. (c) Give the equation for the φ variable and show that it has the solution (2-45) if the separation constant is appropriately defined.

2-8 A particle of mass m_0 is just being bound by a one-dimensional potential well of width $2a$ and depth $-V_0$. What is the minimum value of $V_0 \neq 0$?
(Hint: This problem can be solved with the simple de Broglie wave concept, but this must be justified. Otherwise, the more tedious way of using boundary conditions at $x = a$ and $x = -a$ must be used. In either case ask yourself what the condition of "just being bound" implies.)

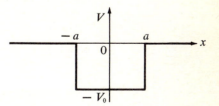

2-9 A parallel beam of particles of kinetic energy T_0 is sent over a potential drop as shown in the figure on p. 70. Calculate the reflection coefficient of the particles at the potential step as a function of T_0 and V_0.

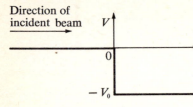

Direction of
incident beam

2-10 (a) Calculate the energies of the lowest two $l = 0$ states of a particle (mass m_0) in a closed spherical box of radius R and give the normalized wave functions for these states. (b) Calculate the energies in Mev if m_0 = mass of a neutron, $R = 5$ F.

2-11 Suppose N identical particles are placed into a closed cubical box so that only one particle is in one state and that the states are filled in order of increasing energy. (a) Establish a relation between N and $(n_x^2 + n_y^2 + n_z^2)_{max}$. (Hint: Each set of integral numbers n_x, n_y, n_z represents one state. Hence, if one plots the occupied states as points in a Cartesian coordinate system with axes along n_x, n_y, n_z the volume of the occupied space is just equal to the total number of occupied states.) (b) What is the maximum energy to which the states are filled? (This entire problem is difficult.)

2-12 A beam of particles of rest mass m_0 and kinetic energy T encounters a potential step of height V_0 ($>T$) as shown. Calculate (a) The wave function for $x > 0$. (b) The reflection coefficient at the potential step. (Answer without calculation, if you wish.) (c) Is the wave function for $x > 0$ changed if the potential step is decreased to zero at $x > L$? (d) Does the reflection coefficient at $x = 0$ change under this condition?

2-13 (a) Show that for a spherical potential $V(r)$ such that $V(r \to \infty) = 0$, the radial solution of Schrödinger's equation, Eq. (2-47), for a bound state has the asymptotic solution

$$u(r \to \infty) \sim e^{-\kappa r}$$

where κ is given by

$$\tfrac{1}{2}\hbar^2 \kappa^2 / m_0 = |E|$$

(b) Also show that Eq. (2-47) is satisfied for $r \to 0$ by the solution

$$R(r \to 0) \sim r^l$$

2-14 (a) Use Fig. 2-8 to estimate the overall energy (in Mev) which is released if U^{238}

fissions spontaneously into two *equal* fragments with the release of four neutrons, and if the fission products decay to stable end-product nuclei (to which Fig. 2-8 applies).   (b) What fraction of the U^{238} mass is converted into energy?

2-15 Refer to Appendix C.   (a) What is the total binding energy of Ca^{40} in Mev and what is the average binding energy per nucleon?   (b) The total electronic binding energy of an atom of atomic number Z is approximately given by $15.73\ Z^{\frac{7}{3}}$ ev. What corrections (in percent) must be made to the answers of part (a), if we wish to compute the purely nuclear total and average binding energies?

2-16 Calculate, from the semiempirical mass formula, the binding energies of the last neutron in Pb^{207} and Pb^{208}. Use any set of consistent energy parameters.

2-17 Use the semiempirical mass formula to determine which isobars with $A = 102$ should be stable and compute the most stable Z [Eq. (2-134)]. Use any consistent set of energy parameters. See Appendix C for the actual stable isobars.

2-18 Use the semiempirical mass formula to compute the mass difference between $_{29}Cu^{64}$ and $_{30}Zn^{64}$. (Compare with Fig. 4-28.)

2-19 (a) Use the semiempirical mass formula to compute the decay energy in alpha decay. (The alpha-decay energy is equal to the negative of the alpha separation energy.)   (b) Apply your equation to the decay of $_{84}Po^{212}$ to $_{82}Pb^{208}$ using any consistent set of energy parameters. (Compare with Fig. 4-17.) Use the empirical value of 28.3 Mev for the binding energy of the alpha particle.

2-20 (a) Use the semiempirical mass formula to compute, for a given A, the relation between Z and N for a nucleus which has zero proton separation energy. (b) Compute N/Z for the case $A = 100$, using a consistent set of energy parameters. (See Fig. 4-11.)

2-21 Use the semiempirical mass formula to calculate the percentage contribution to the average binding energy per nucleon of the volume energy, the surface energy, the coulomb energy, and the asymmetry energy for $A = 60$ and 240. (See Fig. 2-16.)

2-22 On the basis of the single-particle shell model, including spin-orbit coupling, what would be the expected ground-state spectroscopic configurations of the following nuclei:

$$C^{11}, Sc^{45}, Ni^{61}, Ge^{73}, In^{109}, Ta^{181}, Tl^{203}, Am^{241} ?$$

The measured spins are, respectively,

$$\tfrac{3}{2}, \tfrac{7}{2}, \tfrac{3}{2}, \tfrac{9}{2}, \tfrac{9}{2}, \tfrac{7}{2}, \tfrac{1}{2}, \tfrac{5}{2}$$

If you find any discrepancy between your expectations and experiment, try to give some explanation.

2-23 In odd-odd nuclei, an interaction between the last odd neutron and the last odd proton must be taken into account in order to explain ground state spins. The coupling favors parallel intrinsic spins of the odd proton and odd neutron. (This is not to be confused with the pairing coupling for identical nucleons which favors anti-parallel orbital angular momenta.) On this basis give the expected ground-state spins and parities for the following nuclei:

$$N^{14}, Cl^{38}, Y^{90}, Bi^{206}$$

The measured spins are, respectively,

$$1, 2, 2, 6$$

2-24 The energies (in Mev) and spins of the lowest excited states of Ta^{182} are

Energy, Mev	Spin
0.100	2
0.329	4
0.680	6

(a) Do these values agree with the rotational model for permanently deformed nuclei? (b) What is the moment of inertia of the nucleus (in g-cm^2) about the axis of rotation?

2-25 Use Fig. 2-30 to compute the density of levels of Sc^{41} as a function of excitation energy. Break up the excitation-energy scale into 1-Mev intervals and compute the number of levels per Mev. By making a plot on semilog paper, check whether Eq. (2-164) is approximately satisfied. If so, compute the constants ρ_0 and a.

2-26 The following mirror nuclei decay by emitting positrons whose maximum energy in Mev is listed:

C^{11}	0.97	P^{29}	3.95
N^{13}	1.18	S^{31}	4.40
O^{15}	1.73	A^{35}	4.96
F^{17}	1.75	Ca^{39}	5.49
Mg^{23}	3.09	Sc^{41}	5.95
Al^{25}	3.24		

Make a plot similar to Fig. 2-33 and determine the radius constant R_0 assuming each nucleus is a uniformly charged sphere. [See Eq. (4-122) for the relation between positron decay energy and mass difference.]

INTERACTIONS
OF NUCLEAR RADIATIONS ⬛⬛⬛⬛⬛⬛⬛⬛⬛⬛ 3
WITH MATTER

3-1 INTRODUCTION

The chapters following the present one will deal with the dynamic properties of nuclei: radioactive decay and nuclear reactions. The present chapter is concerned with the experimental investigation of these properties. This always requires the detection of nuclear particles or electromagnetic radiation, briefly called *nuclear radiations*. Usually the intensity (number of detected events per unit time) and the (kinetic) energy of the radiation are determined. Most intensity measurements rely on ionization produced when the radiations pass through matter. The energy measurement either involves ionization or atomic excitation, or the deflection of charged particles in electric and magnetic fields. For low-energy neutrons and gamma rays, crystal diffraction also gives a very precise energy determination.

Strictly speaking, these subjects lie outside the field of nuclear physics. Yet they are so basic to nuclear investigations that we must mention at least the most important concepts. For more extensive discussions the reader is referred to other books.[1] The passage through matter of charged particles, neutrons and gamma radiation, will be treated separately because they each involve different characteristic processes.

3-2 INTERACTION OF CHARGED PARTICLES WITH MATTER

A charged particle passing through neutral atoms interacts mainly by means of the coulomb force with the electrons in the atoms. Even though in each encounter the particle loses on the average not more than a few electron volts of kinetic energy, ionization and excitation of atoms give the greatest energy loss per unit path length of the particle. The loss of kinetic energy in a nuclear encounter would be much larger, but such collisions are extremely rare compared to atomic encounters, roughly in proportion to the area of cross section of a nucleus compared to that of an atom, i.e., 10^{-24} cm²/10^{-16} cm² $= 10^{-8}$. Hence, they do not contribute appreciably to the overall energy loss.

For kinetic energies larger than about M_0c^2, where M_0 is the rest mass of the particle, energy loss by emission of electromagnetic radiation becomes increasingly important. The radiation is called *bremsstrahlung* (decelerating radiation). It is caused by the same mechanism as the emission of continuous x rays. The basic process can be understood classically. According to Maxwell's equations, any accelerated charge radiates electromagnetic radiation (see Sec. 4-4b). If a charged particle passes close to a nucleus, its velocity vector will be rapidly changed (at least in direction if not in magnitude), so that the particle undergoes an acceleration and hence it radiates.

A crude idea of the important concepts of the energy loss process by collision (i.e., excitation and ionization of atoms) can be obtained by assuming that the charged particle is heavy and that it collides with a free electron. The loss of kinetic energy of the heavy particle must then be equal to the gain of kinetic energy of the electron. The latter can be estimated from the impact given to the electron as the charged particle passes by. If x and y coordinates are chosen as shown on Fig. 3-1, the impact equations for the electron are

$$\int F_x \, dt \approx 0 \qquad\qquad (3\text{-}1)$$

$$\int F_y \, dt = p_e \qquad\qquad (3\text{-}2)$$

where $\mathbf{F} = \mathbf{F}_x + \mathbf{F}_y$ = coulomb force exerted on the electron
     t = time
     p_e = momentum imparted to electron (only the y component is nonzero)

[1] Burcham, 1963, chaps. 5–8; or Evans, 1955, chaps. 18–28.

Equation (3-1) is a good approximation because the velocity of the heavy particle is practically unaffected by the encounter. If the *impact parameter*[1] of the collision is b, the integral in Eq. (3-2) can be estimated from the *time of the impact*

$$\Delta t \approx \frac{b}{v} \qquad (3\text{-}3)$$

where v is the speed of the heavy particle, and the average magnitude of F_y during that time

$$(F_y)_{\text{ave}} \approx \frac{ze^2}{b^2} \qquad (3\text{-}4)$$

where ze is the charge of the heavy particle. Hence from Eq. (3-2)

$$p_e \approx \frac{ze^2}{bv} \qquad (3\text{-}5)$$

FIGURE 3-1   Encounter between a heavy charged particle of mass M_0 and a free electron of mass m_0. The impact parameter b is indicated.

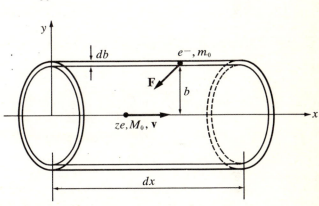

A much cleaner estimate of the momentum imparted to the electron [Eq. (3-5)] can be obtained by an application of Gauss' law of electrostatics

$$\int \mathbf{E} \cdot d\mathbf{S} = 4\pi q \qquad \text{(in electrostatic units)} \qquad (3\text{-}6)$$

where $\mathbf{E}$ is the electric field at the surface of any closed volume surrounding a charge q, and $d\mathbf{S}$ is an element of surface. Applying this theorem in a system of reference in which the heavy particle is at rest to an infinite cylinder of radius b, as shown in Fig. 3-1, we find

$$\int \frac{F_y}{e} 2\pi b \, dx = 4\pi ze \qquad (3\text{-}7)$$

---

[1] The impact parameter in a collision between two particles is the smallest distance at which the centers of the particles would pass each other, if there were no force between the particles.

With respect to the heavy particle, the electron travels a distance $dx = v\,dt$ along the cylinder surface in a time dt so that

$$\int F_y\,dt = 2\frac{ze^2}{bv}$$

$$= p_e \tag{3-8}$$

in accordance with Eq. (3-2). This yields the more correct estimate of p_e

$$p_e = 2\frac{ze^2}{bv} \tag{3-9}$$

so that the kinetic energy gained by the electron (mass m_0) and lost by the heavy particle is

$$\frac{p_e^2}{2m_0} = \frac{2z^2e^4}{m_0b^2v^2} \tag{3-10}$$

If there are n atoms per unit volume, each with Z electrons, then in a path length dx there will be

$$nZ \cdot 2\pi b\,db \cdot dx \tag{3-11}$$

electrons within a distance b to $b + db$ from the path of the heavy particle. To each of these electrons, the particle loses an amount of energy given by expression (3-10) so that the total energy loss per unit path is

$$-\frac{dT}{dx} = \int_{b_{min}}^{b_{max}} nZ\,2\pi b\,db\,\frac{2z^2e^4}{m_0b^2v^2}$$

$$= \frac{4\pi e^4z^2nZ}{m_0v^2}\ln\frac{b_{max}}{b_{min}} \tag{3-12}$$

This expression is approximate, because for nearly head-on collisions Eq. (3-10) is not valid.[1] Nevertheless, it suffices for the present purpose.

In reality the electrons in the stopping material are not free but bound to atoms (or to the solid if the material is in solid form). Since each atom has electronic energy levels the particle cannot transfer *any* energy to the atom unless it excites the atom at least to its first excited state. In classical terms, we can argue that the time of impact [Eq. (3-3)] must not be longer than the period of rotation of a typical electron in its orbit in order that energy will be transferred to the electron in the atom.

$$\Delta t_{max} \approx \frac{1}{\nu} \tag{3-13}$$

[1] A complete derivation can be found in Evans, 1955, chap. 18.

where v is the frequency of rotation. Using Eq. (3-3) this gives

$$b_{max} \approx \frac{v}{\nu} \qquad (3\text{-}14)$$

The minimum impact parameter is limited by the uncertainty principle, because the electron cannot be located with respect to the heavy particle more closely than its de Broglie wavelength. If Eq. (2-93) is applied in the relative coordinate system of the electron and the particle, we see that

$$b_{min} \approx \frac{h}{m_0 v} \qquad (3\text{-}15)$$

Therefore, we obtain

$$-\frac{dT}{dx} \approx \frac{4\pi e^4 z^2 nZ}{m_0 v^2} \ln \frac{2 m_0 v^2}{I_{ave}} \qquad (3\text{-}16)$$

where $h\nu$ has been replaced by a mean ionization and excitation potential I_{ave} of the atoms in the stopping material[1] and a factor 2 has been added in the logarithm term in accord with an exact quantum mechanical calculation.[2] The other symbols in this equation are, as previously,

ze = charge of heavy particle

m_0 = mass of electron

v = speed of heavy particle

nZ = number of electrons per unit volume in stopping material

Experimentally, the energy loss is determined by the number of *ion pairs* formed along the path of the particle. By ion pair, we describe the positive and negative constituents which result from an ionizing encounter. If on the average, an energy w has to be lost by the heavy particle in order to produce one ion pair, the number of ion pairs i per unit path is given by

$$-\frac{dT}{dx} = wi \qquad (3\text{-}17)$$

The quantity w is the result of complicated processes: (1) there is atomic excitation as well as ionization, (2) an ejected electron can have sufficient energy [see Eq. (3-10)] to produce secondary ionization in turn.[3] Empirically, in a given

---

[1] Theoretically and experimentally, $I_{ave} \sim Z$, approximately (Bloch, 1933). The proportionality constant is about 13 ev for most substances, except H_2 and He, for which it is 19 and 22 ev, respectively (Sternheimer, 1961).

[2] For $v \approx c$, the terms $-\ln(1 - v^2/c^2) - v^2/c^2$ have to be added to the logarithm term in Eq. (3-16).

[3] This can be seen quite clearly on Fig. 3-5a. See also Prob. 3-6. The ejected electrons are often called *delta rays*.

material, w is, fortuitously, practically independent of the kinetic energy or of the nature of the particle. For air, it has the value 35.0 ev for 5-kev electrons, 35.2 ev for 5.3-Mev alpha particles, and 33.3 ev for 340-Mev protons.

Before comparing expression (3-16) with experiment, it is convenient to introduce the concept of mean range $\bar{R}$, which is the average distance traveled

FIGURE 3-2 Empirical energy loss curve for protons in air. (*a*) Low energy portion. Eq. (3-16) follows this curve down to $T \approx 0.3$ Mev with $I_{ave} = 80$ ev. Below this energy, capture and loss of electrons reduces the average value of z (see Fig. 3-3). (*b*) High energy portion. The minimum energy loss occurs at $T \approx 1500$ Mev. (By permission from H. A. Bethe and J. Ashkin, in E. Segrè, (ed.), "Experimental Nuclear Physics," vol. 1, John Wiley & Sons, Inc., New York, 1953, as adapted by Burcham, 1963.)

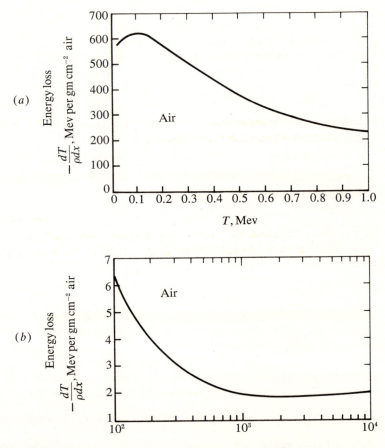

(in a given direction) before the particle loses its kinetic energy T_0 completely:

$$\bar{R} = \int_0^R dx = \int_{T_0}^0 \left|\frac{dx}{dT}\right| dT$$

$$= \int_0^{T_0} \left(-\frac{dT}{dx}\right)^{-1} dT \tag{3-18}$$

If the logarithm term in Eq. (3-16) were independent of v

$$\bar{R} \sim \int_0^{T_0} T \, dT \sim T_0^2 \tag{3-19}$$

but actually no such simple range curve is found. (See Figs. 3-9 and 3-10.) The range of a single particle may be slightly larger or smaller than expression (3-18) because there are statistical variations in the amount of energy lost per unit path length and in the total number of ions pairs formed. This is called *straggling.*

Figure 3-2 gives an experimentally determined energy loss curve for protons in air. Since the number of atoms per unit volume n is related to density of the material ρ, Avogadro's number $\mathscr{N}$, and the atomic weight $\mathscr{A}$ by

$$n = \frac{\mathscr{N}\rho}{\mathscr{A}} \tag{3-20}$$

the energy loss is often given as $-dT/(\rho \, dx)$ because it is then independent of the physical constitution of the stopping material.

Curve (a) of Fig. 3-2 shows that at very low energies the energy loss of the proton decreases rather than increases as might be expected from the $1/v^2$ dependence of Eq. (3-16). This results from intermittent capture and subsequent loss of electrons by the proton, which decreases the average value of z in Eq. (3-16) with decreasing speed v of the proton (Fig. 3-3). As the proton slows down, the probability of electron capture predominates over the loss until the proton finally becomes a hydrogen atom. It can then lose energy only by *elastic collisions.* The number of ion pairs per unit path towards the end of the particle's track shows this same reduction caused by the capture and loss process (Fig. 3-4).

The cloud-chamber photographs shown in Fig. 3-5 illustrate these effects very clearly (for an alpha particle). In a cloud chamber supersaturated vapor, usually water, is artificially generated. The ion pairs produced by a charged particle serve as condensation centers for the vapor. By proper timing and illumination, it is possible to photograph tracks whose droplet density is directly proportional to i [Eq. (3-17)].[1] In Fig. 3-5a the ionization produced by the ejected electrons, called *delta rays,* can be seen. As the alpha particle loses energy the delta rays become less energetic. Towards the end of the range (Fig. 3-5d) the ionization first increases noticeably, as shown in Fig. 3-3, and then

---

[1] For more details on cloud chambers, see Burcham, 1963, sec. 6-2.

FIGURE 3-3 Mean charge z of slow protons and alpha particles as a function of their speed v. (By permission from Evans, 1955.)

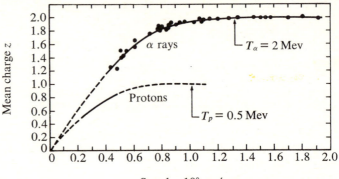

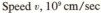

Speed v, 10^9 cm/sec

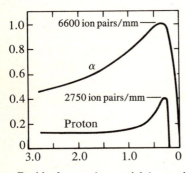

Residual range (α-particle), cm air

FIGURE 3-4 Number of ion pairs per unit path for a single proton and a single alpha particle as a function of residual range. The residual range is the distance left to travel until the particle comes to rest. The horizontal scale is such that on the left part of the diagram both particles have similar speeds. The proton range then is 0.2 air-cm shorter than the alpha-particle range. (By permission from Evans, 1955.)

falls off as the mean charge of the alpha particle decreases. At the same time, the kinks in the alpha-particle tracks show the occurrence of atomic collisions with the cloud-chamber gas.

The energy loss of fast electrons in matter is caused by the same mechanisms as for heavy charged particles, and the energy loss formula is practically identical to Eq. (3-16). The following differences, however, should be noted: (1) Because the incident electron and the electron in the stopping material have the same mass, there is much more scattering of the incident electron (in fact we cannot tell which one was the incident electron). Hence, the path length in the stopping material can be considerably longer than the straight line distance or range in the material. Figure 3-6 illustrates this situation.[1] (2) As the charge z of

---

[1] Note that immediately after collision, the angle between the electrons is 90°. This is a consequence of the nonrelativistic laws of energy and momentum conservation [see Eq. (3-29)], applicable here because the incident electron has only 56-kev energy.

the incident particle never changes, the energy loss of electrons by ionization is appreciable down to energies in the ev range. This causes the blobs at the end of the electron tracks in Fig. 3-6. (3) For a given kinetic energy, electrons have higher speeds than heavy particles. Therefore, the energy loss by radiation for electrons is important at much lower energies than for protons, for example. This is shown in Fig. 3-7.

FIGURE 3-5 Alpha-particle tracks in a cloud chamber. On each figure the initial energy and the equivalent path length in air at 1 atm and 15°C is shown. Note the more energetic delta rays at the higher energies. [By permission from T. Alper, On Delta Rays and the Relation between Range and Velocity of Slow Electrons, *Z. Physik*, **76**: 172 (1932), J. Springer, Publishers, Berlin. Reproduced from W. Gentner, H. Maier-Leibnitz, and W. Bothe, "An Atlas of Typical Expansion Chamber Photographs," Pergamon Press, London, 1954.]

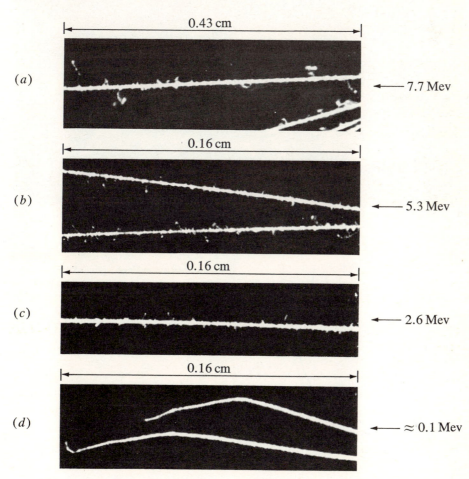

(a) 0.43 cm   7.7 Mev

(b) 0.16 cm   5.3 Mev

(c) 0.16 cm   2.6 Mev

(d) 0.16 cm   ≈ 0.1 Mev

FIGURE 3-6 Track of a 56-kev electron in a cloud chamber. The arrow points to the start of the electron track. In this picture, 59-kev x-rays were used to release the initial electron causing the small blob at the beginning of the track because of an Auger effect. This is further described at end of Sec. 3-4c. [By permission from L. H. Martin, J. C. Bower, and T. H. Laby, *Proc. Roy. Soc.* (*London*) **A148:** 40 (1935). Reproduced from W. Gentner, H. Maier-Leibnitz, and W. Bothe, "An Atlas of Typical Expansion Chamber Photographs," Pergamon Press, London, 1954.]

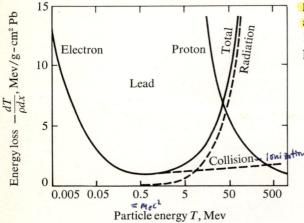

FIGURE 3-7 Energy loss of electrons and protons in lead. (From W. Heitler, 1954, as adapted by Burcham, 1963, by permission.)

A typical experimental arrangement for the detection of charged particles is presented in Fig. 3-8a. In a *good-geometry* situation, collimators are used which prevent particles, scattered by the absorber, from reaching the detector. Curves (b), (c), and (d) give a schematic representation of the absorption curves for heavy charged particles, monoenergetic electrons, and beta rays, respectively. Beta rays have a distribution of initial energies (see Sec. 4-6d) and hence have a particularly complicated absorption curve.[1]

FIGURE 3-8 Schematic absorption curves for charged particles emitted by a radio-active source. (*a*) Experimental arrangement. (*b*) Absorption curve for heavy particles. (*c*) Absorption curve for monoenergetic electrons. (*d*) Absorption curve for beta rays.

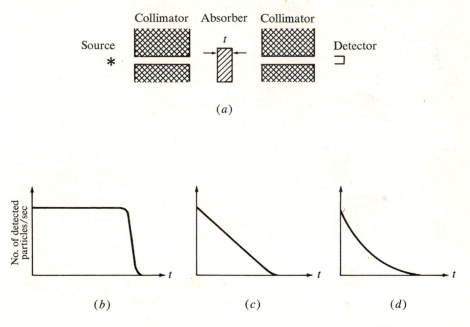

The range of charged particles, as determined by an absorption curve, or in a cloud chamber or photographic emulsion,[2] is a convenient but not very precise method of determining the energy of a charged particle. The method is limited because of straggling to an accuracy between 1 and 5 percent. Figure 3-9 shows a range-energy curve for protons in air at atmospheric pressure; in

---

[1] Empirically it is found that the initial part of the absorption curve follows approximately $e^{-(constant)t}$, where t is the thickness of the absorber and the constant depends on the beta-ray end-point energy.

[2] In a photographic emulsion, the passage of ionizing radiation through the silver halide crystals causes developable grains to be formed. For details see Powell, Fowler, and Perkins, 1959.

FIGURE 3-9 Range-energy relationship for protons in air at 1 atm and 15°C. The relationship of Eq. (3-19) is shown in dotted lines. (Adapted from H. A. Bethe and J. Ashkin, Passage of Radiations through Matter, in E. Segrè (ed.), "Experimental Nuclear Physics," vol. 1, John Wiley & Sons, Inc., New York, 1953.)

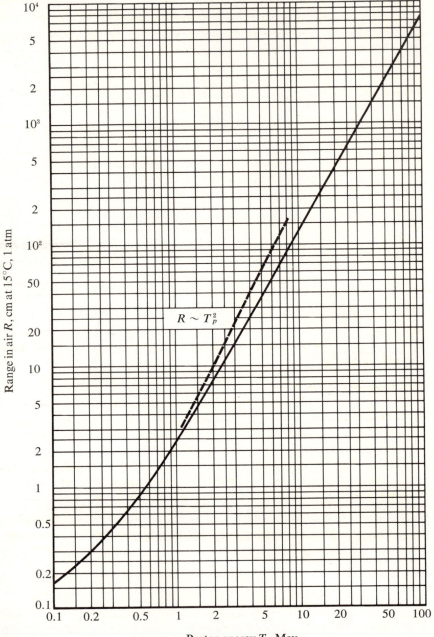

$R \sim T_p^2$

Proton energy T_p, Mev

Range in air R, cm at 15°C, 1 atm

aluminum, a similar curve is obtained with the range decreased by a factor of approximately[1] 1/1600. A range-energy curve for electrons is given in Fig. 3-10. Neither curve has the simple energy dependence of Eq. (3-19).

More accurate methods of energy determination consist in measuring electronically the total number of ion pairs formed in the stopping material (ionization chamber, gas proportional counter, solid-state proportional counter) or in detecting the total light emitted in the ionization-excitation process (scintillation counter).[2] In all these cases one relies on the approximate constancy of w in Eq. (3-17). The detectors must always be calibrated with particles of known energy.

The energy of charged particles can be determined absolutely from their path in known electric or magnetic fields (in an evacuated space, to avoid scattering). In the simplest of such instruments, called spectrometers,[3] the charged particles are deflected into a circular path of radius r by a uniform magnetic field of induction B. The momentum p of the particle is then given by

$$p = eBr \qquad\qquad (3\text{-}21)$$

(all electromagnetic or MKS units) and the kinetic energy can be determined from Eqs. (2-9) and (2-10) or from $\frac{1}{2}p^2/m_0$ in the nonrelativistic situation.

3-3   INTERACTION OF NEUTRONS WITH MATTER

The interaction of neutrons with matter is not only of experimental or theoretical interest but has important practical applications, particularly in the operation of reactors.[4] The present discussion will be restricted to the energy loss of neutrons by elastic collisions.

3-3a Energy loss of neutrons.   Because neutrons are neutral, they cannot lose energy by ionization. Nuclear encounters, though rare, are the only possible means of energy loss. Most of these collisions are elastic; that is, the struck nucleus is not excited, but in some cases the inelastic excitation can contribute to the energy loss.

In any collision, momentum is conserved, and if the collision is elastic, kinetic energy is conserved also. We will neglect relativistic effects, i.e., our derivation will not be applicable to neutrons of energy above 200 Mev.[5] It is convenient to consider the collision both in the laboratory (lab.) and in the center-of-mass (c.m.) coordinate systems (Fig. 3-11).

Before collision, particle 1 of mass M_1 and speed v_1 collides with particle 2

---

[1] Evans, 1955, chap. 22, sec. 3e.
[2] For details see Burcham, 1963, sec. 6-1.
[3] For details see Burcham, 1963, sec. 7-2.
[4] Kaplan, 1962, chap. 18; Segrè, 1964, chap. 12.
[5] The derivation is also applicable to elastic collisions of electrons of energy below about 100 kev, as in Fig. 3-6. For the relativistic case, see Segrè, 1964, app. G.

FIGURE 3-10 Range-energy relationship for electrons in aluminum. To obtain the range in cm, divide the range in mg/cm² by the density of Al, 2700 mg/cm³. The relationship of Eq. (3-19) is shown in dotted lines. [Empirical relationship of L. Katz and A. S. Penfold, *Rev. Mod. Phys.* **24**: 28 (1952) as shown by J. B. Marion, *Nucl. Data Tables, U.S. At. Energy Comm. 1960*, part 3, available from U.S. Government Printing Office, Washington, D. C.]

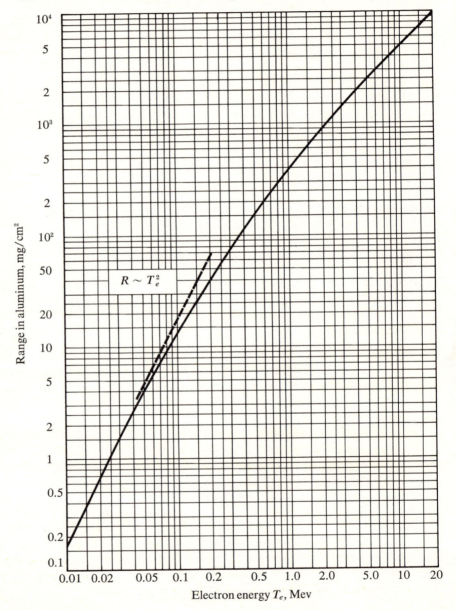

Range in aluminum, mg/cm²

Electron energy T_e, Mev

$R \sim T_e^2$

of mass M_2, which is at rest. After the collision, the particles move with speeds v_1' and v_2', respectively, at angles θ_1 and θ_2 in the lab. system (Fig. 3-11a). The collision takes place in a plane because initially there is no momentum perpendicular to the velocity $\mathbf{v}_1$. The center of mass of the system moves with a speed

$$v_0 = \frac{M_1 v_1}{M_1 + M_2} \tag{3-22}$$

in the direction of $\mathbf{v}_1$.

FIGURE 3-11   Elastic collision of two particles.   (a) Laboratory system.   (b) Center of mass system.   (c) Laboratory system.

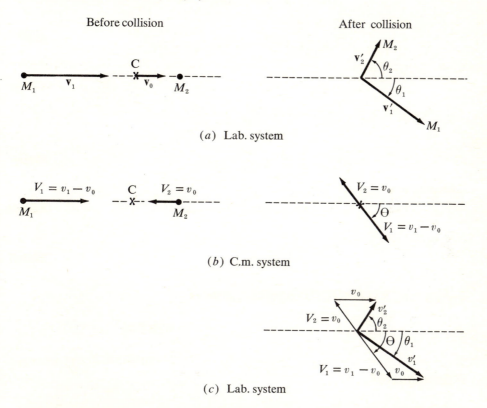

(a) Lab. system

(b) C.m. system

(c) Lab. system

If the vector $\mathbf{v}_0$ is subtracted from all velocity vectors, the center of mass evidently is at rest, so that the c.m. system is obtained by this procedure (Fig. 3-11b). In this system both particles have identical and opposing vector momenta, both before and after collision. In addition, conservation of kinetic energy requires that each particle retain its speed during the collision.[1] Hence, if V_1 and

[1] See Prob. 3-14.

V_2 are the c.m. speeds of the two particles before collision, we have

$$M_1 V_1 = M_2 V_2 \tag{3-23}$$

and V_1 and V_2 are also the speeds after the collision. Furthermore

$$V_2 = v_0 \tag{3-24}$$

by construction of the c.m. system. The angle of collision Θ in the c.m. system depends on the details of the collision.

To return to the lab. system, the velocity $\mathbf{v}_0$ has to be added to all velocity vectors in the c.m. system (Fig. 3-11c). All relations concerning speeds and angles can now be read off the figure immediately. For example, the energy of particle 1 in the lab. system after collision is

$$T_1' = \tfrac{1}{2} M_1 v_1'^2$$
$$= \tfrac{1}{2} M_1 (V_1^2 + v_0^2 + 2 V_1 v_0 \cos \Theta) \tag{3-25}$$

This has maximum and minimum values at $\Theta = 0°$ and $\Theta = 180°$, respectively:

$$(T_1')_{\max} = T_1 \tag{3-26}$$

$$(T_1')_{\min} = \tfrac{1}{2} M_1 (V_1 - v_0)^2$$
$$= \tfrac{1}{2} M_1 (v_1 - 2v_0)^2$$
$$= T_1 \left(\frac{M_1 - M_2}{M_1 + M_2} \right)^2 \tag{3-27}$$

If $M_1 = M_2$, we find from Eq. (3-25)

$$T_1' = T_1 (1 + \cos \Theta)/2, \qquad (T_1')_{\min} = 0 \tag{3-28}$$

and from the fact that $V_1 = V_2 = v_0$ it is easy to show that

$$\theta_1 + \theta_2 = 90° \tag{3-29}$$

for any collision.

3-3b Energy distribution of neutrons after collision.   For neutrons up to several Mev kinetic energy, it is approximately correct to assume in a collision with a nucleus the distribution of the scattered neutrons is isotropic in the c.m. system.[1] Under these conditions the number of neutrons scattered into a given solid angle $d\Omega$ in the c.m. system is proportional to $d\Omega$. The probability of scattering into $d\Omega$ is thus

$$P(d\Omega) = \frac{d\Omega}{4\pi}$$

$$= \frac{2\pi \sin \Theta \, d\Theta}{4\pi}$$

$$= \tfrac{1}{2} \sin \Theta \, d\Theta \tag{3-30}$$

----

[1] This is another way of stating that the neutron elastic scattering cross section in the c.m. system is isotropic. See Sec. 5-4d and Appendix A-2.

Each neutron which is scattered into an angular interval Θ to $\Theta + d\Theta$ also has its energy changed from T_1 to the interval T_1' to $T_1' + dT_1'$, where dT_1' is given by differentiating Eq. (3-25)

$$dT_1' = (-)M_1 V_1 v_0 \sin \Theta \, d\Theta \tag{3-31}$$

Hence, the probability of scattering into this energy interval is

$$P(dT_1') = P(d\Omega)$$

$$= \frac{dT_1'}{2M_1 V_1 v_0} \tag{3-32}$$

from Eqs. (3-30) and (3-31). Figure 3-12 shows a plot of the probability distribution $P(dT_1')/dT_1'$, which is of course just the energy distribution of the neutrons after a single collision.

FIGURE 3-12 Energy distribution of neutrons after one collision.

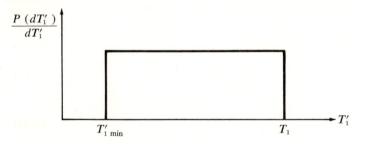

For a hydrogen scatterer, $(T_1')_{\min} = 0$ [Eq. (3-28)], so that the average energy after collision is

$$(T_1')_{\text{ave}} = \tfrac{1}{2}T_1$$

The energy distribution of neutrons after n collisions can also be calculated.[1] We might expect that after n collisions the average energy is approximately

$$(T_1')_{\text{ave}} \approx (\tfrac{1}{2})^n T_1$$

If neutrons collide with a hydrogen scatterer, the recoil protons have the same energy distribution as the scattered neutrons. Figure 3-13 shows an actual proton recoil energy distribution in an organic scintillator. As we mentioned at the end of Sec. 3-2, in proton energy loss by ionization and excitation, the number of photons emitted happens to be nearly proportional to the energy loss. If the light emitted by a scintillator is allowed to fall on the photosensitive surface of a photomultiplier, the energy loss in the scintillator can be determined by electronic means. The rounding off of the energy distribution in Fig. 3-13, as

[1] Segrè, 1964, chap. 12.

FIGURE 3-13 Pulse-height spectra of recoil protons in an organic scintillator (stilbene) produced by monoenergetic neutrons from the $H^2(d,n)$ reaction. (a) Experimental spectra taken with a stilbene crystal 1 cm in diameter and 0.1 cm thick. (b) Proton energy spectra deduced from the top figure, after subtracting background and correcting for nonlinear response of the stilbene scintillator. [By permission from C. D. Swartz and G. Owen, Recoil Detection in Scintillators, in J. B. Marion and J. L. Fowler, (eds.), "Fast Neutron Physics," vol. 1, chap. IIB, Interscience Publishers, Inc., New York, 1960.]

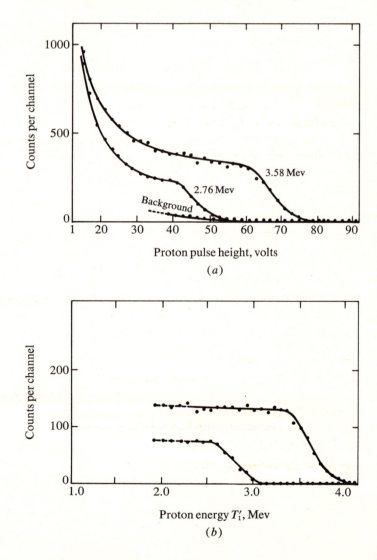

compared to the ideal distribution of Fig. 3-12, is caused by statistical effects in the photomultiplier. If the scintillator is calibrated with neutrons of known energy, it can be used to determine unknown neutron energies. This is generally successful only if there are not more than a few monoenergetic neutron groups, well separated in energy.

Absolute neutron energy measurements can be made by determining the flight time of neutrons over a known distance. Using electronic timing techniques with resolutions of 10^{-9} sec and path lengths of meters, this method can be extended to the Mev energy range.[1] The diffraction of neutrons by crystals is also used for energy determination below a few ev, where the de Broglie wavelength of neutrons [Eq. (2-11)] becomes comparable to crystal lattice spacings (10^{-8} cm $= 10^5$ F).

3-4 INTERACTION OF GAMMA RADIATION WITH MATTER

Gamma radiation is the name commonly given to electromagnetic radiation of nuclear origin. It usually has a wavelength smaller than about 10^5 F or photon energy larger than about 0.1 Mev [see Eq. (2-2)]. The interaction processes of gamma rays with matter are complex. Certain features of the interaction can be understood with classical arguments, i.e., on the basis of Maxwell's equations; but only quantum electrodynamics describes the correct physical situation. Indeed we should recall that the photoelectric effect and the Compton effect are two of the basic experiments which first demonstrated the inadequacy of Maxwell's equations and the need to introduce quantum concepts (see Sec. 2-2a).

Classical electrodynamics shows that an accelerated charged particle radiates. Hence, when electromagnetic radiation of frequency v encounters a loosely bound electron, the induced acceleration causes the electron to reradiate some of the electromagnetic energy at the same frequency. The phenomenon is called Thomson scattering; its quantum extension is the Compton effect.

Next imagine that the electron is bound to an atom and circulates around a nucleus with a frequency v_0. As with the forced oscillation of any resonating system, we expect the most pronounced effect of the incident electromagnetic radiation on the electron to occur if $v = v_0$. The largest energy transfer to the electron occurs under this *resonance* condition,[2] and the electron then has the greatest chance to be separated from the atom. The quantum extension of this process is the photoelectric effect.

A third interaction mechanism, pair production by gamma rays, has no classical analog.

3-4a Attenuation of gamma rays.   The attenuation of a beam of gamma rays through an absorber is fundamentally different from that of a beam of heavy

---

[1] The time of flight of a 1-Mev neutron is 0.7×10^{-9} sec/cm of path. See Burcham, 1963, sec. 7-3, for details.

[2] This reasoning formed the basis of Eq. (3-13).

charged particles (Fig. 3-8b). The latter undergo many small interactions which hardly affect the direction of the particle. If gamma rays pass through matter, each gamma ray either does not interact at all, or it is removed completely from the beam by absorption or scattering.[1] This causes an exponential attenuation with increasing absorber thickness.

FIGURE 3-14   Attenuation of gamma-ray beam by an absorber.   (a) Intensity of beam at various points.   (b) Attenuation curve.

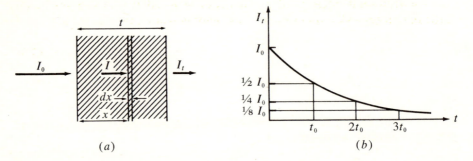

(a)   (b)

Suppose I_0 gamma rays per unit time are incident normally on an absorber and that at a penetration depth x the unaffected beam has an intensity I, as shown in Fig. 3-14. The fractional number of gamma rays removed from the beam will be proportional to dx, because the individual attenuation processes are completely independent of each other.

$$-\frac{dI}{I} = \mu \, dx \qquad (3\text{-}33)$$

where the proportionality constant μ is called the *linear attenuation coefficient*.[2] Integration of Eq. (3-33) for an absorber of thickness t gives

$$I_t = I_0 e^{-\mu t} \qquad (3\text{-}34)$$

It should be noted that I_t is the intensity of the unaffected beam. The thickness t_0 needed to attenuate the beam to one-half of its original intensity is called *half-value thickness.* Substitution into Eq. (3-34) yields

$$t_0 = \frac{\ln 2}{\mu} = \frac{0.693}{\mu} \qquad (3\text{-}35)$$

For 3-Mev gamma rays attenuated by lead, $t_0 \approx 0.5$ in.

Since attenuation is caused by three independent processes, Compton effect, photoelectric effect, and pair production, we can write

$$\mu = \mu_{\text{C}} + \mu_{\text{E}} + \mu_{\text{P}} \qquad (3\text{-}36)$$

[1] The scattering of gamma rays into very small (forward) angles by bound electrons in atoms, called Rayleigh scattering, is ignored here. See Evans, 1955, chap. 23; Burcham, 1963, sec. 5-4.
[2] In common usage μ is also called the total linear absorption coefficient, although scattering and absorption both contribute to attenuation of the gamma-ray beam.

where each partial attenuation coefficient is proportional to the probability of occurrence of the particular process. Each coefficient is also proportional to the number of atoms per unit volume of absorber [see Eq. (3-20)], and thus it is useful to define the *mass absorption coefficient* μ/ρ where ρ is the density of the material. The mass absorption coefficient is independent of the physical state of the absorber. In terms of it we can rewrite Eq. (3-34)

$$I_t = I_0 e^{-(\mu/\rho)\rho t} \tag{3-37}$$

The fractional intensity of gamma rays removed from the beam by one particular process only, for example, by Compton scattering, is equal to

$$\frac{\mu_C}{\mu} \frac{I_0 - I_t}{I_0} = \frac{\mu_C}{\mu} (1 - e^{-\mu t}) \tag{3-38}$$

The exponential term is not $e^{-\mu_C t}$, because all attenuation processes occur even if only Compton scattered gamma rays are observed.

Typical mass attenuation coefficients are shown in Figs. 3-15 and 3-16. The various interaction mechanisms are predominant over different gamma-ray energy regions. Their energy dependence cannot be understood without complex quantum mechanical calculations.[1] We will, however, discuss the general nature of each process.

3-4b Compton effect. It is easy to show that energy and momentum cannot be conserved if a photon is completely absorbed by a free electron at rest.[2] Therefore, in the interaction of a gamma ray with a loosely bound electron, the gamma ray must be scattered (with an appropriate loss in energy). This corresponds to the classical reemission of electromagnetic radiation mentioned previously.

Figure 3-17 shows the interaction process and our notation. From conservation of momentum we find

$$p_r = p_r' \cos \theta + p_e \cos \varphi \tag{3-39}$$

$$0 = -p_r' \sin \theta + p_e \sin \varphi \tag{3-40}$$

Conservation of energy gives

$$E_r = E_r' + T_e \tag{3-41}$$

Upon elimination of p_e and φ we find that

$$\lambda' - \lambda = \frac{h}{m_0 c} (1 - \cos \theta) \tag{3-42}$$

where λ' and λ are the wavelengths of the scattered and incident gamma rays, respectively. In Eqs. (3-39) to (3-41), expressions (2-1) and (2-3) must be used

[1] See Heitler, 1954.
[2] Electrons in matter are of course neither free nor at rest. Nevertheless, they can be considered as such for the present purpose if (1) the incident photon energy is much larger than the binding energy of the electrons (ionization potential in a gas or work function in a solid), and (2) the incident photon momentum is much larger than the momentum of the struck electron.

for E_r and p_r and the relativistic expressions (2-8) and (2-9) for p_e and T_e. The quantity h/m_0c is called the *Compton wavelength* of the electron and has the value

$$\frac{h}{m_0c} = 2{,}426 \text{ F} \tag{3-43}$$

The shift in wavelength (3-42) is independent of the incident gamma-ray energy.

FIGURE 3-15 Mass attenuation coefficient for gamma rays in aluminum as a function of gamma-ray energy. The coefficients for the photoelectric effect (μ_E/ρ), the Compton effect (μ_C/ρ), and the pair effect (μ_P/ρ) are shown separately. To obtain the coefficients in cm^{-1}, multiply by the density of Al = 2.70 g/cm^3. (By permission from Evans, 1955.)

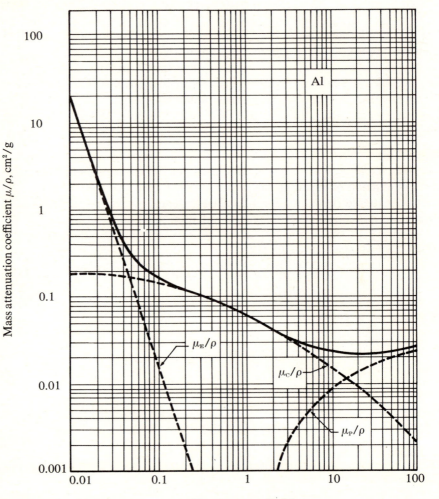

For very high-energy gamma rays, i.e., $E_r \gg m_0 c^2$ ($= 0.511$ Mev) or $\lambda \ll h/m_0 c$, we can neglect λ with respect to λ' for all scattering angles θ except near $0°$, so that

$$E_r' \approx \frac{m_0 c^2}{1 - \cos \theta} \tag{3-44}$$

FIGURE 3-16 Mass attenuation coefficient for gamma rays in lead as a function of gamma-ray energy. The coefficients for the photoelectric effect (μ_E/ρ), the Compton effect (μ_C/ρ), and the pair effect (μ_P/ρ) are shown separately. To obtain the coefficients in cm^{-1}, multiply by the density of Pb $= 11.35$ gm/cm^3. (By permission from Evans, 1955.)

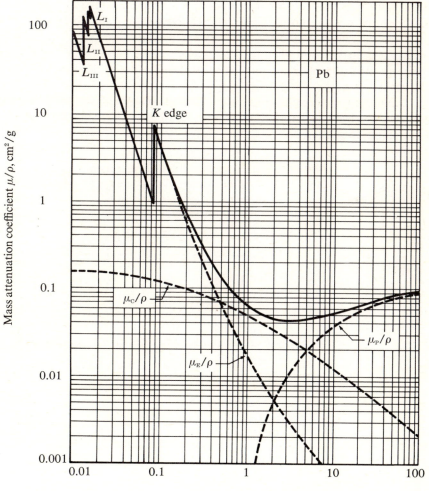

FIGURE 3-17 Interaction of gamma ray with a free electron.

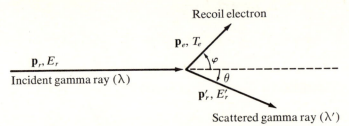

The probability of Compton scattering can be calculated only with the Dirac equation. The angular distribution of Compton scattered gamma radiation, i.e., relative number of photons scattered into a small angle $d\Omega$ at an angle θ, is shown in Fig. 3-18. For very small photon energies, it approaches the classical distribution (Thomson scattering). One can show[1] from Maxwell's equations that the classical angular distribution should be proportional to $1 + \cos^2 \theta$.

Gamma-ray detectors are sensitive to the ionization produced by the recoil electrons. Typical calculated recoil energy distributions are shown in Fig. 3-19. If the scattered photon has minimum energy ($\theta = 180°$ in Fig. 3-17), the recoil

FIGURE 3-18 Angular distribution (intensity per unit solid angle) of Compton-scattered gamma rays as a function of the scattering angle for various incident gamma-ray energies E_r. All curves have been normalized at 0°. (By permission from Heitler, 1954.)

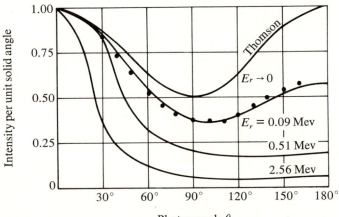

electron will have maximum energy

$$T_e(\text{max}) = E_r - E'_r(\theta = 180°) \qquad (3\text{-}45)$$

For high-energy gamma rays, according to Eq. (3-44)

$$T_e(\text{max}) \approx E_r - \tfrac{1}{2}m_0 c^2$$
$$\approx E_r - 0.255 \text{ Mev} \qquad (3\text{-}46)$$

We can understand that the recoil energy distribution peaks near $T_e(\text{max})$ (see Fig. 3-19), because for an appreciable range of angles θ near $\theta = 180°$, $\cos\theta$

FIGURE 3-19 Energy distribution of Compton electrons as a function of electron energy for various incident gamma-ray energies E_r. [By permission from C. M. Davisson and R. D. Evans, *Rev. Mod. Phys.*, **24**: 79 (1952).]

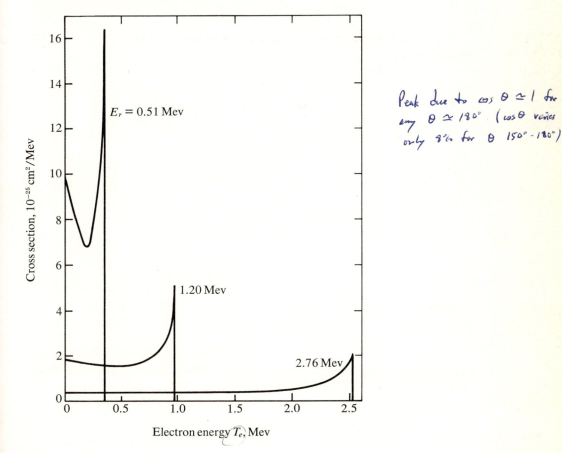

Peak due to $\cos\theta \simeq 1$ for any $\theta \simeq 180°$ ($\cos\theta$ varies only 8% for θ $150° - 180°$)

remains close to the value −1 and so E_r' remains close to the value $\frac{1}{2}m_0c^2$. Figure 3-20 shows an actual Compton-electron energy distribution from the gamma-ray bombardment of an organic scintillator. The rounding off of the energy distribution is caused by statistical effects in the photomultiplier used to detect the light from the scintillator.

FIGURE 3-20   Pulse-height spectra of Compton electrons produced by 0.51- and 1.28-Mev gamma rays (from Na^{22}) in an organic scintillator (stilbene). No correction has been made for the slightly nonlinear energy response of the scintillator. The stilbene crystal was in the form of a cylinder of 3.8-cm diam and 2-cm length. (By permission from C. D. Swartz and G. Owen, Recoil Detection in Scintillators, in J. B. Marion and J. L. Fowler, (eds.), "Fast Neutron Physics," vol. 1, chap. IIB, Interscience Publishers, Inc., New York, 1960.)

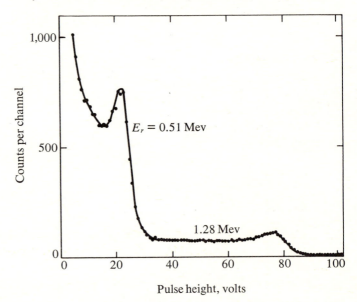

Pulse height, volts

If the gamma ray is scattered by a bound electron which is not removed from its atom, Eqs. (3-39) to (3-41) also hold, but p_e and T_e now refer to the entire atom which recoils with the bound electron. Hence in Eq. (3-42), m_0 must be replaced by the mass of the atom. The shift in wavelength is negligible for most purposes. This type of scattering is called *Rayleigh scattering*. It increases with the atomic number Z of the scatterer, since the binding energy of the inner electrons is proportional to Z^2 so that an increasing fraction of the atomic electrons must be considered as bound. The angular distribution is *not* like that given in Fig. 3-18, because the radiation scattered from all bound electrons in one atom interferes coherently. As a result, Rayleigh scattering is sharply peaked about $\theta = 0°$.

Gamma radiation can also scatter on the nucleus without excitation (Thomson scattering) or with excitation. The first process interferes coherently with Rayleigh scattering, but has a much smaller probability.

FIGURE 3-21   Interaction of a gamma ray with a bound electron.

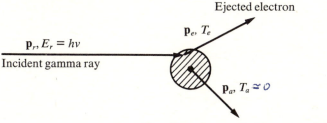

Ejected electron

$\mathbf{p}_e, T_e$

$\mathbf{p}_r, E_r = h\nu$

Incident gamma ray

$\mathbf{p}_a, T_a \approx 0$

Recoiling atom, excited to energy state E_B

3-4c Photoelectric effect.   A gamma ray can transfer its energy to an electron originally bound in an atom because the atom can then take up some of the recoil momentum, as shown in Fig. 3-21. Conservation of momentum

$$\mathbf{p}_r = \mathbf{p}_e + \mathbf{p}_a \tag{3-47}$$

and conservation of energy

$$E_r = T_e + T_a + E_B \tag{3-48}$$

can be simultaneously satisfied. In Eq. (3-48), E_B is the binding energy of the electron in the atom,[1] which is also the excitation energy of the atom after the electron has been ejected. It is not difficult to show that the recoil kinetic energy T_a is of order $(m_0/M_0)T_e$, where m_0 and M_0 are the masses of the electron and atom, respectively. Since $m_0/M_0 \approx 10^{-4}$, T_a can be neglected for most purposes so that

$$T_e = h\nu - E_B \tag{3-49}$$

For gamma rays above about $\frac{1}{2}$ Mev, photoelectrons are most probably ejected from the K shell of an atom because, for these electrons, the classical resonance condition (Sec. 3-4) is most nearly satisfied. The angular distribution of photoelectrons is given in Fig. 3-22. For low-energy gamma rays, the distribution is practically symmetric about $\theta = 90°$; this can be understood classically since the photoelectrons should be ejected parallel to the electric field vector of the incident radiation.

The probability of photoelectric emission is shown on Figs. 3-15 and 3-16. The probability increases as $h\nu \to E_B$ or $\nu \to \nu_0$, where $\nu_0 = E_B/h$ is the frequency of the *absorption edge*. This is caused by the resonance effect mentioned

[1] For K electrons, $E_B \approx 13.6 (Z-1)^2$ ev.

FIGURE 3-22   Angular distribution (intensity per unit solid angle) of photoelectrons as a function of the angle between the electrons and the incident gamma rays. The energy of the incident gamma radiation is given for each curve. [By permission from C. M. Davisson and R. D. Evans, *Rev. Mod. Phys.*, **24**: 79 (1952).]

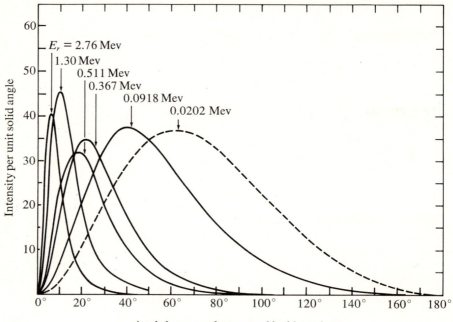

Angle between electron and incident photon

in Sec. 3-4. Above the absorption edge, the probability of photoelectric emission varies roughly as $Z^5 E_r^{-3.5}$, where Z is the atomic number of the interacting atom.

The photoelectric effect is always accompanied by a secondary process, since the atom does not remain in its excited energy state E_B. Either x rays are emitted by the atom, or electrons are released from the outer atomic shells, which carry away the available excitation energy. These are called *Auger electrons*.[1] In any dense material, the secondary radiations are absorbed in turn with a high probability. This occurs in most scintillators which are used for gamma-ray detection (see Fig. 3-26).

3-4d Pair production.   Crudely speaking, the Schrödinger equation is the quantum mechanical equivalent of the nonrelativistic conservation of energy

---

[1] In Fig. 3-6 the initial electron was a photoelectron from the K shell of an argon atom, released by an incident 59-kev photon. Instead of K x rays from the argon atom, a 3-kev Auger electron was ejected from the L shell, causing the small blob at the start of the photoelectron track.

equation [Eq. (2-28)] and the Dirac equation is the quantum mechanical equivalent of the relativistic equation

$$W = \pm(p^2c^2 + m_0^2c^4)^{\frac{1}{2}} + V \qquad (3\text{-}50)$$

which can be obtained from Eq. (2-10). For the moment let us assume that the potential V is equal to zero. The ambiguity in sign of the square root in Eq. (3-50) is not a mathematical accident. It was shown by Dirac that positive energies W represent a particle of rest mass m_0 and momentum $\mathbf{p}$ and that the negative energy states represent a particle of rest mass $-m_0$ and momentum $-\mathbf{p}$ (Fig. 3-23).

FIGURE 3-23 Creation of an electron-positron pair according to the Dirac theory.

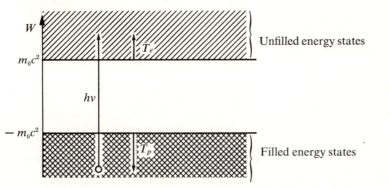

Since the minimum value of p^2 is zero, no particles can occupy the energy interval $m_0c^2 > W > -m_0c^2$. To overcome the difficulty that ordinary (positive-energy) electrons would make transitions to negative energy states until they are completely filled, Dirac assumed that nature is such that (1) all negative energy states are filled with electrons in the absence of any field or matter, and (2) no effect of these electrons is noticeable in the absence of any field or matter.[1] Suppose, now, that an electron is ejected from a negative energy state by action of a gamma ray. The creation of the *hole* in the negative energy states means that the system under consideration acquires a mass $-(-m_0)$, a momentum $-(-\mathbf{p})$, and a charge $-(-e)$. The creation of the hole, therefore, corresponds to the appearance of a particle of mass m_0, momentum $\mathbf{p}$, and charge $+e$. The kinetic energy T_p of this *hole particle* would be equal to $\frac{1}{2}p^2/m_0$ in the non-relativistic approximation or be given accurately by Eq. (2-9). The particle is called a *positron.* It was discovered by Anderson (1932). It occurs in several nuclear processes and its existence is well established.

---

[1] More complex formulations of the Dirac equation avoid these drastic assumptions but give identical results.

When the hole is created, an electron also appears in a positive energy state with kinetic energy T_e. From conservation of energy (see Fig. 3-23)

$$h\nu = T_e + T_p + 2m_0c^2 \tag{3-51}$$

It is possible to show that this equation cannot be satisfied together with the momentum conservation $\mathbf{p}_r = \mathbf{p}_e + \mathbf{p}_p$. Therefore an electron-positron pair can be produced only in the neighborhood of a third particle which can take up some momentum. If the third particle is a nucleus, it will take up hardly any energy (as in the photoelectric effect), so that Eq. (3-51) still holds to a very good approximation. The minimum energy for pair production obviously occurs for $T_e + T_p = 0$, that is, $h\nu = 2m_0c^2 \approx 1.02$ Mev.

Figure 3-24 shows a set of stereoscopic photographs of an electron-positron pair produced in an expansion cloud chamber by a 7-Mev photon. The cloud chamber was filled with air. A magnetic field was applied in order to determine the momenta of both particles. In this particular event the momenta happened to be nearly equal.

FIGURE 3-24 Electron-positron pair produced by 7-Mev gamma ray in a cloud chamber (stereoscopic photograph). The cloud chamber was filled with air at 1.75 atm pressure. A magnetic field was applied to curve the particle tracks. The energy loss of the particles can be noticed by the decreasing radius of the spiral paths. The gamma-ray source was external to the chamber. Note the Compton and photoelectrons released in the chamber wall. [By permission from J. A. Phillips and P. G. Kruger, *Phys. Rev.* **76**: 1471 (1949). Reproduced from W. Gentner, H. Maier-Leibnitz, and W. Bothe, "An Atlas of Typical Expansion Chamber Photographs," Pergamon Press, London, 1954.]

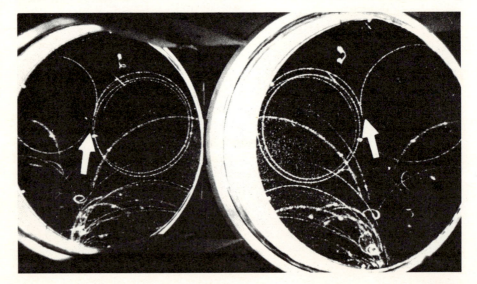

The angular and energy distributions of the electrons and positrons can be calculated from the Dirac theory. The energy distribution is shown in Fig. 3-25. Roughly speaking, all energies are equally probable. The total probability of pair formation, corresponding to the area under the curves of Fig. 3-25, is given in Figs. 3-15 and 3-16 for Al and Pb. The probability is roughly proportional to Z^2. Pair formation can also occur in the neighborhood of atomic electrons if $h\nu \geq 4m_0c^2$. It is less probable than pair production near a nucleus by a factor approximately equal to $1/(4Z)$.[1]

[1] Evans, 1955, chap. 24, sec. 2h.

FIGURE 3-25 Energy distribution of positrons (or electrons) as a function of the relative positron energy $T_p/(T_e + T_p)$ for various incident gamma-ray energies E_r. Distributions for pair formation in aluminum and lead are shown. [By permission from C. M. Davisson and R. D. Evans, *Rev. Mod. Phys.* **24**: 79 (1952).]

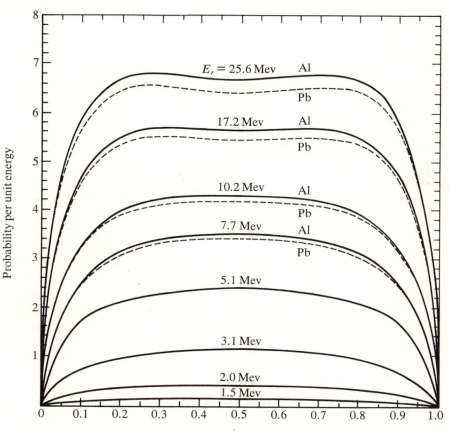

3-5 INTERACTION OF POSITRONS WITH MATTER

The energy loss of positrons passing through matter, as that of electrons, occurs by ionization and bremsstrahlung (see Sec. 3-2). In addition, positrons can *annihilate* with electrons by a process which is the inverse of pair production. The annihilation probability turns out to be greatest for very slow positrons. If such a positron annihilates with a free electron, conservation of linear momentum requires that at least two gamma rays be emitted so that each of these has an energy equal to $m_0c^2 (= 0.511$ Mev). This radiation is called *annihilation radiation*. If the electron is bound in an atom, annihilation with the production of a single photon can occur because the atom can take up the necessary momentum. This is a relatively rare process.

A positron and an electron can also form a type of atom, in which each one of the two particles moves about its common center of mass. This structure is called *positronium* and was first detected by Deutsch (1949–1951). The positronium atom is short-lived ($\approx 10^{-10}$ sec or $\approx 10^{-7}$ sec, depending on the relative spin orientations of the two particles), because the electron and positron annihilate each other.[1]

3-6 DETECTION OF NUCLEAR RADIATIONS

We discuss briefly the *inorganic scintillator* and the *semiconductor* detectors, both of which have advanced the art of gamma-ray and charged-particle detection tremendously.[2]

The most common inorganic scintillator detector, invented by Hofstadter (1949), uses a sodium iodide crystal. The light emitted in the ionization-excitation process lies in the ultraviolet and is not easily detected. The crystal is therefore activated with a fraction of a percent of thallium iodide which shifts the wavelengths into the visible region, suitable for photomultiplier detection. In the sodium iodide crystal, an entering gamma ray interacts mainly by the three processes we have discussed: Compton effect, photoelectric effect,[3] and, if $E_r > 1.02$ Mev, pair production. Figure 3-26 shows a typical pulse height spectrum for 1.37- and 2.75-Mev photons in a sodium iodide detector. The Compton-electron distributions are very similar to those shown in Fig. 3-20 for an organic scintillator, in which photoelectric effect and pair production are usually negligible. The photoelectric peaks are not only due to the true photoelectric effect, but also due to the reabsorption of scattered low-energy Compton gamma rays. The peaks are spread out in energy because of statistical effects in the photomultiplier. Annihilation radiation can also be reabsorbed, but the

---

[1] For further details see Deutsch and Berko, 1965.

[2] For other detectors see Burcham, 1963, chap. 6.

[3] Since the photoelectric effect and pair production are roughly proportional to Z^5 and Z^2, respectively, these two interactions take place predominately in the iodine atoms.

FIGURE 3-26 Pulse-height spectra of 1.37- and 2.75-Mev gamma rays (from Na²⁴) in a sodium iodide detector. The photo peaks are marked by the energy only. For the other peaks, the following notation is used: P2 = two-escape pair peak, P1 = one-escape pair peak, C = Compton edge. The detector was a cylinder, 7.6-cm diam and 7.6-cm long, placed at a distance of 10 cm from the radioactive source. A Be absorber removed beta radiation. Na²⁴ emits 1.37- and 2.75-Mev gamma rays of equal intensity. (By permission from R. L. Heath, "Scintillation Spectrometry Gamma-Ray Spectrum Catalog," Phillips Petroleum Company, Idaho Falls, 1964.)

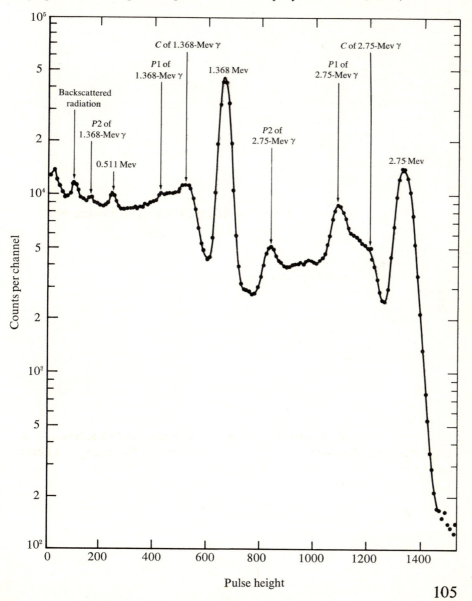

FIGURE 3-27 Pulse-height spectra of 1.37- and 2.75-Mev gamma rays of Na²⁴ in a lithium-drifted germanium detector. The same notation as in Fig. 3-26 applies. The detector had a 1.9-cm diam and was 0.5 cm thick. (By permission from A. J. Tavendale, "Proceedings of the Second International Symposium on Nuclear Electronics, Paris, Nov. 1963," p. 235, European Nuclear Energy Agency, Paris, 1964.)

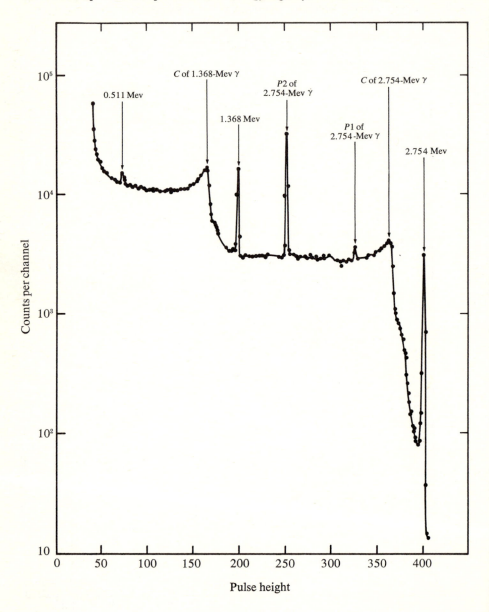

Table 3-1 Common nuclear detectors

Particle	Detector	Method of detection	Remarks
Heavy charged particles; electrons	Ionization chamber & proportional counter	Total no. of ion pairs determined by collecting charged partners of one sign, e.g. electrons.	Can be used to determine T_0 if particle stops in chamber.†
	Semiconductor detector	Ionization produces electron-hole pairs. Total charge is collected.	Used to determine T_0.
	Geiger counter	Ionization initiates brief discharge.	Good for intensity determination only.
	Cloud chamber or photographic emulsion	Path made visible by ionization causing droplet condensation or developable grains.	Can be used to determine T_0 from range. Type of particle can be recognized from droplet or grain count along path.
	Scintillation detector	Uses light produced in excitation of atoms.	T_0 proportional to light produced.
Neutrons	Any of the above using proton recoils from thin organic lining or nuclear reactions with appropriate filling gas	Ionization by recoiling protons.	T_0 from end point of recoil distribution.
		Ionization by reaction products.	Some reactions can be used to determine T_0.
	Organic scintillator and photomultiplier	Using light produced in excitation of atom by recoiling protons.	T_0 from end point of recoil distribution.
			The above methods for neutrons require $T_0 > 0.1$ Mev.
Gamma rays	Geiger counter	Electrons released in wall of counter ionize gas and initiate discharge.	Good for intensity determination only.
	NaI scintillation detector	Light produced in ionization and excitation by electrons released in the three interaction processes.	T_e proportional to light produced; $h\nu$ inferred from electron energy distributions.
	Semiconductor detector	Electrons produced create electron-hole pairs. Total charge is collected.	$h\nu$ inferred from electron energy distributions.

† The initial kinetic energy of the particle is called T_0.

finite probability of escape of either one photon or both means that often either the energy $E_r - m_0c^2$ or the energy $E_r - 2m_0c^2$ is deposited in the crystal. This gives rise to the *one-escape* and *two-escape* pair peaks.

In a semiconductor detector, the electrons released in the excitation process are detected directly by electronic means. The semiconducting material is treated so that its normal conductivity is practically zero, making it sensitive to the extra electrons produced by an entering gamma ray or charged particle. Because of the electronic structure of a semiconductor, only about 3 ev are needed to produce a conducting electron in a typical semiconducting material (silicon or germanium).[1] This reduces statistical effects in the number of charge carriers, giving energy resolutions down to 0.1 percent. Figure 3-27 shows a semiconductor spectrum for the same gamma radiation as in Fig. 3-26. There is an appreciable increase in resolution. As in a sodium iodide detector, the electron distributions from the various interaction processes are apparent.

The most common nuclear detectors are listed in Table 3-1. Considerable sophisticated electronics has been developed to determine such phenomena as the coincident occurrence of several radiations or the total charge collected by a detector.[2] In the detection of nuclear particles, fluctuations caused by the random nature of the disintegration or production processes must be considered.[3]

PROBLEMS

3-1   The energy loss of a heavy charged particle (charge ze) of speed v in a material with n atoms per unit volume (atomic number Z) is, in electrostatic cgs units,

$$-dT/dx = [4\pi e^4 z^2 nZ/(m_0 v^2)][\ln (2m_0 v^2/I) - \ln (1 - v^2/c^2) - v^2/c^2]$$

where m_0 = mass of an electron.
Show that this expression passes through a minimum as v is varied and find the approximate kinetic energy of the particle at that speed. Relativistic expressions must be used.

3-2   Omitting the last terms $[-\ln (1 - v^2/c^2) - v^2/c^2]$ in the energy-loss expression in Problem 3-1, calculate the energy loss of a 10-Mev alpha particle in aluminum, for which $I = 150$ ev.

3-3   (a) If the energy loss of a 10-Mev proton in air is 50 kev/cm, what is the energy loss of a 40-Mev alpha particle? (The answer can be written immediately.)

[1] For details see Dearnaley and Northrup, 1963.
[2] The technology of nuclear particle detection is rapidly evolving. A fairly complete account is given in Siegbahn, 1965. Recent developments are summarized in the *Trans. Nucl. Sc.* of the Institute of Electrical and Electronics Engineers.
[3] See Sec. 4-2a, or for more details, Evans, 1955, chaps. 26–28.

(b) Assuming the energy-loss equations for protons and for nonrelativistic electrons are identical, at what kinetic energy does an electron have the same energy loss as a 10-Mev proton?

3-4 The energy-loss formula for heavy charged particles in a monoatomic substance is often written

$$-dT/dx = 4\pi e^4 z^2 nZB_e/(m_0 v^2)$$

where B_e is called the atomic stopping number per electron.

Suppose an absorbing material consists of a fraction f_1 (by number) of atoms of kind 1 (Z_1, A_1) and a fraction f_2 of atoms of kind 2 (Z_2, A_2). **(a)** Derive an equation for the energy loss of heavy charged particles in this material in terms of B_{e1} and B_{e2}. Call the mass density of the material ρ. **(b)** For 8-Mev alpha particles, observed values of B_e are 5.6 for hydrogen and 4.0 for nitrogen. Compute the energy loss of 8-Mev alpha particles in ammonia gas (NH_3) at NTP.

3-5 Compute the number of ion pairs per millimeter of path generated by 2-Mev protons in nitrogen gas at NTP. Assume $I = 80$ ev and $w = 35$ ev.

3-6 **(a)** Show that the number of *delta rays* per unit path length released by a heavy charged particle passing through matter is given by

$$\frac{2\pi e^4 z^2 nZ}{m_0 v^2} \frac{dT_e}{T_e^2}$$

if the delta rays have kinetic energies between T_e and $T_e + dT_e$. **(b)** Compute the number of delta rays per millimeter of path, which have kinetic energies in excess of 0.5 kev, for 2-Mev protons passing through nitrogen gas at NTP.

3-7 Show that alpha particles and protons of the same initial *speed* have approximately the same range in any stopping material. Why is this not accurately true? Which particle should have a slightly longer range and why?

3-8 What is the energy of a proton which has approximately the same range as a 10-Mev alpha particle?

3-9 Figure 3-10 shows that the range of 0.2-Mev electrons in aluminum is 43 mg/cm². Neglecting the effect of differing values of the average ionization potential I, compute the approximate path length of these electrons in air at 1 atm and 15°C. Assume that the energy loss for nonrelativistic electrons is given by the same equation as for heavy charged particles.

3-10 A beam of monoenergetic neutrons is sent into an ionization chamber, filled with monoatomic gas. The energy distribution of the recoiling atoms is found to have a spread equal to 9.5 percent of the maximum recoil energy. What gas is in the ionization chamber?

3-11 A beam of 2.0-Mev neutrons is scattered by a very thin slab of paraffin [approximately $(CH_2)_n$]. The thickness is so small that no multiple scattering of neutrons takes place, only single scattering. What are the energies of neutrons scattered through 90°? What are the corresponding energies of the recoiling nuclei?

3-12 Prove that if a particle collides elastically with another particle of the same mass which is at rest, after the scattering, the angle between the two particles is 90°.

3-13 A particle of mass M_1 collides elastically with a particle of mass M_2 at rest. Show that, if $M_1 > M_2$, the lab. angle of M_1 after the collision cannot exceed the value $\sin^{-1} M_2/M_1$.

3-14 (a) Show that conservation of kinetic energy and linear momentum during an elastic collision (Fig. 3-11) require that in the c.m. system the speed of each particle before the collision be equal to its speed after collision. (b) Does the relative speed of particle 1 with respect to particle 2 change during an elastic collision (1) in the c.m. system, (2) in the lab. system?

3-15 (a) Compute approximately the number of collisions necessary to bring the average energy of 1-Mev neutrons in a large block of carbon to thermal energies ($\approx 1/40$ ev). (b) Make an order of magnitude estimate for the time involved in slowing down the neutrons, assuming a 10-cm collision mean free path.

3-16 A radioactive source is surrounded with a thin absorber to remove beta rays. The remaining gamma radiation is absorbed in aluminum, with the following

Absorber thickness (cm)	Detected activity (counts/min)	Absorber thickness (cm)	Detected activity (counts/min)
0	3,510	1.0	1,740
0.1	3,180	1.5	1,470
0.2	2,870	2.0	1,280
0.3	2,630	3.0	1,000
0.4	2,430	4.0	790
0.5	2,260	5.0	620
0.6	2,120	6.0	510
0.7	2,000	7.0	400

Analyze these data and find (a) The absorption coefficient of each of the gamma rays and the corresponding half-value thickness. (b) The energy of the gamma rays (use Fig. 3-15). (c) The relative intensity of the gamma rays emitted by the source.

3-17 A pencil beam of gamma rays passes through 2.0 cm of lead. The incident beam consists 30 percent of 0.4-Mev photons and 70 percent of 1.5-Mev photons. What is the fractional intensity of the transmitted beam? For data use Fig. 3-16.

3-18 Suppose a lead absorber is used to measure the attenuation coefficient of gamma rays. A study of Fig. 3-16 will convince you that if the observed mass absorption coefficient lies between 1 and 7 cm²/g or between 0.04 and 0.09 cm²/g, two possible gamma-ray energies could give such a coefficient. How would you propose to remove this ambiguity by absorption experiments alone?

3-19 Suppose a beam of gamma radiation, containing a continuous distribution of photon energies up to 50 Mev, such as might occur in bremsstrahlung, is passed through a very thick absorber of lead. Which gamma-ray energy will emerge with the largest relative intensity?

3-20 A radioactive source emits 1-Mev beta rays and 1-Mev gamma rays. The available detector is sensitive to both beta and gamma rays. How would you propose to make the detector insensitive to beta rays, while allowing most of the gamma rays to be detected?

3-21 A beam of 1.6-Mev gamma rays bombards a gold leaf. Some electrons emerge with 0.7-Mev kinetic energy. Assuming that the electrons were not rescattered nor lost energy by bremsstrahlung, can you tell whether they are photo, Compton, or pair electrons and at which angle to the incident gamma-ray beam they were emitted?

3-22 Prove that a photon cannot transfer all its energy to a free electron.

3-23 Prove that a photon (of energy exceeding $2m_0c^2$) cannot undergo a pair effect in free space.

3-24 Prove that in the photoelectric effect on a free atom (for example, in a monoatomic gas), the recoil energy of the atom is of the order of magnitude $(m_0/M_0)T_e$ where m_0 and M_0 are the rest masses of the electron and atom respectively and T_e is the energy of the photoelectron. Assume that the incident gamma-ray energy is much larger than the binding energy of the electron which is ejected from the atom.

3-25 A sodium iodide detector consisting of a 7-cm cube of material is bombarded with a pencil beam of 2.8-Mev gamma rays normal to one face of the cube. **(a)** What fraction of the gamma rays is detected? **(b)** What fraction of the detected gamma rays appears in the photo peak, the Compton distribution, and the pair peaks, assuming no reabsorption of Compton gamma rays or of annihilation quanta? **(c)** Make a rough estimate of the relative fraction of pair events that appear in the full-energy (photo) peak, in the one-escape peak, and in the two-escape peak. Compare with Fig. 3-26. (Attenuation coefficients for photons in sodium iodide can be found in Evans, 1955, chap. 25, sec. 1. The following data will be helpful: for 0.51-Mev photons, $\mu = 0.33 \text{ cm}^{-1}$; for 2.8-Mev photons, $\mu = 0.135 \text{ cm}^{-1}$, $\mu_E = 2.5 \times 10^{-3} \text{ cm}^{-1}$, $\mu_C = 0.113 \text{ cm}^{-1}$, $\mu_P = 0.020 \text{ cm}^{-1}$.)

3-26 Calculate the ionization energy of positronium.

3-27 In Fig. 3-24, the diameter of the cloud chamber is 30 cm. From the additional information given in the figure caption, can you compute an approximate value for the magnetic field which was applied?

RADIOACTIVE DECAY 🞑🞑🞑🞑🞑🞑🞑🞑🞑🞑🞑🞑🞑🞑🞑🞑 4

4-1 INTRODUCTION

The rest of this book is concerned with a discussion of the dynamic, or time-varying, properties of nuclei: radioactive decay and nuclear reactions. Both of these are characterized by a transition from some initial system to some final system, occurring either spontaneously (radioactive decay) or artificially (nuclear reaction). Hence, from a theoretical point of view there is a strong similarity.

Whether a process occurs spontaneously or only artificially is simply a matter of energetics. If the total energy of the final system is less than that of the initial system, the transition can occur spontaneously. Usually, the larger the energy difference the greater is the transition rate. If the total energy of the final system exceeds that of the initial system, a transition will occur only if some energy is furnished to the initial system.

113

In the present chapter we will deal with radioactive decay. Within several years of the discovery of radioactivity, it was found that the naturally occurring radioactive nuclides emit one or more of three types of radiations,[1] which were differentiated by their penetrability:

Alpha rays—stopped by a sheet of paper

Beta rays—stopped by $\frac{1}{16}$ inch of lead

Gamma rays—can penetrate through several inches of lead

Ingenious and careful experiments were performed by many investigators, which showed that alpha rays are nuclei of He^4, beta rays are electrons and gamma rays are electromagnetic radiation. Each of these decay modes illuminates a different aspect of nuclear structure and will be considered in turn. The time dependence of radioactive decay, however, is common to all modes, and will be discussed first.

4-2 RADIOACTIVITY

The initial radioactive nuclide in any decay mode is called the *parent*, and the (heavy) product nuclide is called the *daughter*. The simplest situation occurs if the daughter is stable. If several successive generations of daughters are radioactive, we speak of a *radioactive decay chain*.

4-2a Decay of a single radioisotope.   The basic experimental fact of radioactive decay is that the probability for any one nucleus to decay within a small time interval dt is independent of any external influence,[2] including the decay of another nucleus. All nuclei of a given nuclide have identical decay probability. Hence the probability $P(dt)$ of a radioactive decay in dt is proportional only to dt if dt is small enough so that $P(dt) \ll 1$. The proportionality constant λ, called the *decay constant*, is different for different nuclides and decay modes. Hence

$$P(dt) = \lambda \, dt \qquad\qquad (4\text{-}1)$$

To calculate the probability that a given nucleus survives for a time interval t divide the time interval t into n equal intervals of duration dt. The probability to survive the first interval is

$$1 - P(dt)$$

to survive the second interval

$$[1 - P(dt)]^2$$

and to survive the nth interval

$$[1 - P(dt)]^n$$

---

[1] For a good historical account, see Rutherford, Chadwick, and Ellis, 1930.

[2] In very special situations, it is possible to change the half-life for electron capture (Sec. 4-6f) of a radioisotope by a few percent by embedding it in various chemical compounds (see, e.g., Cooper, Hollander, and Rasmussen, 1965).

which can be written, using Eq. (4-1)

$$(1 - \lambda\,dt)^n = \left(1 - \frac{\lambda t}{n}\right)^n \xrightarrow[\substack{n \to \infty \\ dt \to 0}]{} e^{-\lambda t} \tag{4-2}$$

This is the survival probability for one nucleus. If N_0 identical nuclei were present initially, the number *most probably* surviving after a time t is

$$N = N_0 e^{-\lambda t} \tag{4-3}$$

This equation can also be derived in a simpler fashion by noting that if N nuclei are present, the number probably decaying in a time dt is

$$-dN = P(dt)N$$
$$= \lambda\,dt N \tag{4-4}$$

since every decay reduces N. Equation (4-2) follows by a simple integration.

Equation (4-4) does not imply that in every time interval dt exactly the same number of nuclei dN of the total number N disintegrates. Only the most probable number dN is calculated, since Eq. (4-1) is a statement about probability. It is easy to show that there must be variations in the time intervals between successive decays. Conversely, in any given time interval there must be fluctuations in the number of nuclei decaying.

Consider a sample of N nuclei. From Eq. (4-4) the average time $\bar{t}$ between successive decays is

$$\bar{t} = \frac{dt}{|dN|} = \frac{1}{\lambda N} \tag{4-5}$$

If $\bar{t}$ is short compared to $1/\lambda$, the probability of one decay occurring in the entire sample in a small interval of time dt is just $dt/\bar{t}$. The probability that there be *no* decay in a finite time interval $t(\ll 1/\lambda)$ is equal to $(1 - dt/\bar{t})^{t/dt}$, following the same reasoning that led to Eq. (4-2). Therefore, the probability for no decay within t, but one decay within the interval between t and $t + dt$ is

$$\left[1 - \frac{dt}{\bar{t}}\right]^{t/dt} \frac{dt}{\bar{t}} \xrightarrow[dt \to 0]{} e^{-t/\bar{t}} \frac{dt}{\bar{t}} \tag{4-6}$$

This gives us the distribution of time intervals between successive decays and shows that small time intervals are more probable than long time intervals.

Conversely, in repeated equal time intervals the number of nuclei decaying in a given sample must have statistical fluctuations about the most probable mean (ignoring the decrease in the size of the sample due to decay). One can show,[1] if the most probable number of disintegrations in a time interval t_1 is N_1, there is roughly a 68 percent chance that the actual number of decays lies between $N_1 - \sqrt{N_1}$ and $N_1 + \sqrt{N_1}$.

_e.g._ $N_1 = 100 \rightarrow$ find 90 to 110 still alive 68% of time

[1] Evans 1955, chap. 26.

The *half-life* $t_{\frac{1}{2}}$ of radioactive decay is the time interval in which the original number of nuclei is reduced to one-half [compare Eq. (3-35)].

$$t_{\frac{1}{2}} = \frac{\ln 2}{\lambda} = \frac{0.693}{\lambda} \tag{4-7}$$

The *mean life* τ is the average time of survival of a radioactive nucleus. From Eqs. (4-3) and (4-4)

$$\tau = \frac{\int t \, dN}{\int dN} = \frac{\int_0^\infty t N_0 e^{-\lambda t} \lambda \, dt}{N_0}$$

$$= \frac{1}{\lambda} \tag{4-8}$$

The decay of each radioactive nucleus is signaled by the emission of a decay particle (alpha, beta, or gamma ray). The number of decay particles (*radiation*) dN_r emitted from a sample of N radioactive nuclei in a time dt is

$$dN_r = -dN = \lambda N \, dt \tag{4-9}$$

from Eq. (4-4). The rate of generation of decay particles is called *activity*. From Eqs. (4-6) and (4-2) the activity is

$$\frac{dN_r}{dt} = \lambda N$$

$$= \lambda N_0 e^{-\lambda t}$$

$$= \left(\frac{dN_r}{dt}\right)_0 e^{-\lambda t} \tag{4-10}$$

Common units of activity are the Curie (1C = 3.70×10^{10} dis/sec)[1] and the Rutherford (1R = 10^6 dis/sec).

If a single nuclide can decay by more than one process, for example by alpha and beta decay, the probability of decay is increased. The probabilities of the individual decay modes are additive since the possibility of alpha decay is independent of the possibility of beta decay. For a sample of N nuclei, the decrease of N in a time dt is caused by both decay modes:

$$-dN = dN_\alpha + dN_\beta$$

$$= \lambda_\alpha N \, dt + \lambda_\beta N \, dt \tag{4-11}$$

By integration

$$N = N_0 e^{-(\lambda_\alpha + \lambda_\beta)t} = N_0 e^{-\lambda_{\text{tot}} t} \tag{4-12}$$

[1] By international convention it has been proposed that the Curie should be abbreviated by Ce or Ci to avoid confusion with the coulomb. We use the older notation since no confusion will occur here.

We call $\lambda_\alpha/\lambda_{tot}$ the branching ratio for alpha decay. The experimental half-life is $(\ln 2)/\lambda_{tot}$.

The alpha activity is equal to

$$\frac{dN_\alpha}{dt} = \lambda_\alpha N$$

$$= \lambda_\alpha N_0 e^{-(\lambda_\alpha + \lambda_\beta)t} \qquad (4\text{-}13)$$

and similarly for the beta activity. The decay is governed by $\lambda_\alpha + \lambda_\beta$ because the nuclide is not prevented from beta decay even if only alpha decay is observed. [Compare Eq. (3-38).]

4-2b Production of a radioisotope by nuclear bombardment.   Suppose that a sample of material is bombarded with neutrons and that a radioisotope is produced at a steady rate Q.[1] The radioisotope decays at a rate $-\lambda N$ where N is the number of radioactive nuclei present. The net rate of change of N is therefore

$$\frac{dN}{dt} = Q - \lambda N \qquad (4\text{-}14)$$

Rearranging,

$$\frac{dN}{Q - \lambda N} = dt$$

or

$$\frac{d(Q - \lambda N)}{Q - \lambda N} = -\lambda \, dt$$

By integration

$$Q - \lambda N = (Q - \lambda N)_{t=0} e^{-\lambda t} \qquad (4\text{-}15)$$

If $N_{t=0} = 0$

$$N = \frac{Q}{\lambda}(1 - e^{-\lambda t}) \qquad (4\text{-}16)$$

As soon as the bombardment stops, the radioisotope decays in accordance with Eq. (4-2). The number of radioactive nuclei as a function of time is plotted in Fig. 4-1. Usually it is not worthwhile to bombard for longer than two to three half-lives since $\frac{3}{4}$ to $\frac{7}{8}$ of the maximum number of radioactive nuclei (Q/λ) is then produced. The activity during bombardment is given by λN, not by $-dN/dt$.

4-2c Production of a radioisotope by a decaying parent.   Suppose a parent 1 decays with a decay constant λ_1 and produces a daughter 2 and radiation r_1, and that the daughter decays in turn with a decay constant λ_2 producing a stable

---

[1] This implies that the bombardment hardly uses up the original material, a good assumption except for bombardments in high-flux reactors.

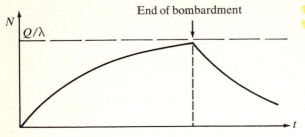

End of bombardment

FIGURE 4-1   Production of radioisotope by nuclear bombardment.

$N_2(0) = 0$

nuclide 3 and radiation r_2.

$$1 \xrightarrow{\lambda_1} 2 + r_1$$
$$\downarrow^{\lambda_2}$$
$$3 + r_2$$

If N_1, N_2, and N_3 are the respective numbers of radioactive nuclei present at any given time t, we find

$$\frac{dN_1}{dt} = -\lambda_1 N_1 \tag{4-17}$$

$$\frac{dN_2}{dt} = \lambda_1 N_1 - \lambda_2 N_2 \tag{4-18}$$

$$\frac{dN_3}{dt} = \lambda_2 N_2 \tag{4-19}$$

In deriving Eq. (4-18) we only need to remember that every decaying nucleus 1 produces one nucleus 2. Since 2 is radioactive, it also decays.

If N_{10} is the original number of nuclei 1 present, we obtain from Eq. (4-17) [see Eq. (4-3)]

$$N_1 = N_{10} e^{-\lambda_1 t} \tag{4-20}$$

Substitution into Eq. (4-18) yields

$$\frac{dN_2}{dt} + \lambda_2 N_2 = \lambda_1 N_{10} e^{-\lambda_1 t} \tag{4-21}$$

The complete solution of this inhomogeneous differential equation consists of a general solution of the homogeneous equation

$$\frac{dN_2}{dt} + \lambda_2 N_2 = 0 \tag{4-22}$$

plus any particular solution of the original equation (4-21). A general solution of Eq. (4-22) is

$$N_2 = C e^{-\lambda_2 t} \tag{4-23}$$

where C is a constant, determined below by initial conditions. A particular

variation of parameters

solution of Eq. (4-21) of the form $N_2 = Ke^{-\lambda_1 t}$ can be attempted. Substitution into Eq. (4-21) then yields $K = N_{10}\lambda_1/(\lambda_2 - \lambda_1)$. The complete solution of Eq. (4-21) is therefore

$$N_2 = \frac{N_{10}\lambda_1}{\lambda_2 - \lambda_1} e^{-\lambda_1 t} + Ce^{-\lambda_2 t} \qquad (4\text{-}24)$$

If initially no nuclei 2 are present we can evaluate C and find

$$N_2 = \frac{N_{10}\lambda_1}{\lambda_2 - \lambda_1} (e^{-\lambda_1 t} - e^{-\lambda_2 t}) \qquad (4\text{-}25)$$

Substitution into Eq. (4-19) allows an immediate integration

$$N_3 = \frac{N_{10}\lambda_1\lambda_2}{\lambda_2 - \lambda_1} \left(\frac{e^{-\lambda_1 t}}{-\lambda_1} - \frac{e^{-\lambda_2 t}}{-\lambda_2} \right) + C' \qquad (4\text{-}26)$$

If $N_3 = 0$ at $t = 0$, the constant C' is determined and

$$N_3 = \frac{N_{10}\lambda_1\lambda_2}{\lambda_2 - \lambda_1} \left(\frac{1 - e^{-\lambda_1 t}}{\lambda_1} - \frac{1 - e^{-\lambda_2 t}}{\lambda_2} \right) \qquad (4\text{-}27)$$

4-2d Special cases. If in the previous example the parent 1 is short-lived compared to the daughter 2, i.e., $\lambda_1 > \lambda_2$, then after a long time $t (\gg 1/\lambda_1)$

$$e^{-\lambda_1 t} \ll e^{-\lambda_2 t}$$

so that Eq. (4-25) gives

$$N_2 \approx \frac{N_{10}\lambda_1}{\lambda_1 - \lambda_2} e^{-\lambda_2 t} \qquad (4\text{-}28)$$

Therefore, the decay of 2 after a long time is determined only by its own half-life. This is shown schematically in Fig. 4-2a.

FIGURE 4-2 Decay of a radioactive daughter. (a) Short-lived parent. (b) Long lived parent.

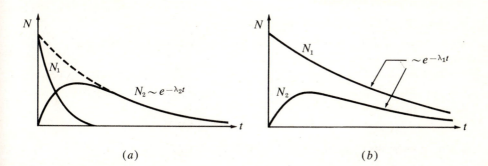

(a)   (b)

If the parent 1 is long-lived compared to the daughter 2, i.e., $\lambda_2 > \lambda_1$, then after a long time $t(\gg 1/\lambda_2)$.

$$N_2 \approx \frac{N_{10}\lambda_1}{\lambda_2 - \lambda_1} e^{-\lambda_1 t} \tag{4-29}$$

The decay of 2 after a long time is therefore determined by the half-life of 1, as shown on Fig. 4-2b. Also, under these conditions,

$$\frac{N_2}{N_1} \approx \frac{\lambda_1}{\lambda_2 - \lambda_1} \tag{4-30}$$

and

$$\frac{\lambda_2 N_2}{\lambda_1 N_1} = \frac{\text{activity of 2}}{\text{activity of 1}}$$

$$\approx \frac{\lambda_2}{\lambda_2 - \lambda_1} \tag{4-31}$$

This state of affairs is called *transient equilibrium*. If $\lambda_1 \ll \lambda_2$, the activities are equal and *secular equilibrium* is obtained.

4-3 WIDTHS OF DECAYING STATES

The spontaneous decay of a nuclear state may be considered from a quantum mechanical point of view. The finite lifetime of the state causes an uncertainty in its energy. In any experiment the energy will be found to have a spread, called *width*, given by Eq. (2-163)

$$\Gamma = \hbar\lambda = \frac{\hbar}{\tau} \tag{4-32}$$

where τ is the mean life of the state [Eq. (4-8)]. This expression will now be derived.

Mean lives of typical nuclear decay processes lie between 10^{-16} sec and 10^{16} years. Even the shortest lifetime is many times longer than the period of a typical nuclear motion which is of the order of the nuclear traversal time (2-144), i.e., approximately 10^{-22} sec. Therefore from this point of view, the nuclear state in question is practically stable and we can attempt to write its wave function in the form (2-16)

$$\Psi(\mathbf{r},t) = \psi(\mathbf{r})e^{-(iW/\hbar)t} \tag{4-33}$$

where $\mathbf{r}$ represents symbolically all the coordinates of the nucleus and W is the total energy (or mass) of the nucleus.

But the nucleus does decay. The probability of finding it in a given volume element must decrease in accordance with Eq. (4-2), which means that Ψ must have the property

$$|\Psi(\mathbf{r},t)|^2 = |\Psi(\mathbf{r}, t = 0)|^2 e^{-\lambda t} \tag{4-34}$$

where λ is the decay constant.

Substitution of Eq. (4-33) into (4-34) shows that W must be an imaginary quantity. If we call its real part E_0, we find by comparing Eqs. (4-33) and (4-34)

$$W = E_0 - \tfrac{1}{2}i\hbar\lambda \qquad (4\text{-}35)$$

that is,

$$\Psi'(\mathbf{r},t) = \psi(\mathbf{r})e^{-i(E_0/\hbar)t - \lambda t/2} \qquad (4\text{-}36)$$

A decaying state is therefore not a state of definite energy E of the form $\psi(\mathbf{r})e^{-i(E/\hbar)t}$. Nevertheless, it can be represented by a superposition of states of slightly different energies E, each with a different amplitude $A(E)$

$$\Psi'(\mathbf{r},t) = \psi(\mathbf{r}) \int_{-\infty}^{\infty} A(E)e^{-i(E/\hbar)t} \, dE \qquad (4\text{-}37)$$

Using the techniques of Fourier analysis, we can show that the energies E are grouped about a mean energy E_0 with a spread of the order of $\hbar\lambda$. Equating Eqs. (4-36) and (4-37), $A(E)$ is computed from the relation

$$e^{-\lambda t/2} = \int_{-\infty}^{\infty} A(E)e^{-i[(E-E_0)/\hbar]t} \, dE \qquad (4\text{-}38)$$

According to the Fourier theorem[1] any well-behaved function $f(t)$ can be represented as

$$f(t) = \frac{1}{2\pi} \lim_{\Omega \to \infty} \int_{-\Omega}^{\Omega} e^{-i\omega t} \, d\omega \int_{-\infty}^{\infty} e^{i\omega t'} f(t') \, dt' \qquad (4\text{-}39)$$

Applying this to the function $e^{-\lambda t/2 - i(E_0/\hbar)t}$, we find

$$A(E) = \frac{1}{2\pi\hbar} \int_0^{\infty} e^{[i(E-E_0)/\hbar - \lambda/2]t'} \, dt'$$

$$= \frac{i}{2\pi} \frac{1}{E - E_0 + i\hbar\lambda/2} \qquad (4\text{-}40)$$

where it has been assumed that the decaying system was prepared at the time $t = 0$. The probability of finding the system with a given energy E is proportional to the absolute square of the amplitude $A(E)$

$$|A(E)|^2 = \frac{1}{4\pi^2} \frac{1}{(E - E_0)^2 + (\hbar\lambda/2)^2} \qquad (4\text{-}41)$$

This curve, shown in Fig. 4-3, has a *Lorentzian shape*. It peaks at the mean energy E_0 and has a width Γ at half-height given by Eq. (4-32). Substituting numerical values we obtain

$$\Gamma \text{ (in ev)} = 0.66 \times 10^{-15}/\tau \text{ (in sec)}$$

$$= 0.46 \times 10^{-15}/t_{\frac{1}{2}} \text{ (in sec)} \qquad (4\text{-}42)$$

where $t_{\frac{1}{2}}$ is the half-life of the state.

[1] See Franklin, 1940, sec. 299.

FIGURE 4-3 Probability of finding a decaying state with a
definite energy E. The state has a width $\Gamma = \hbar\lambda$ where λ is the
decay constant.

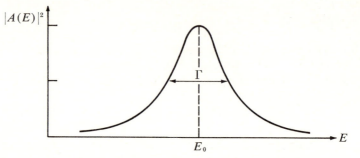

4-4 GAMMA DECAY

A nucleus can be brought to an excited state by a variety of means. For example,
alpha and beta decay can leave nuclei excited, and in many nuclear reactions
excited nuclei are produced.

An excited nucleus can always decay to a lower-energy state by emission
of electromagnetic radiation or by internal conversion (discussed in Sec. 4-4e).
In the simplest case, in which both of the levels involved are single-proton
states,[1] the decay consists of a transition of the proton from the higher to the
lower state. This is analogous to the transition of an excited electron in an atom
from a higher to a lower level, which is accompanied by electromagnetic
emission or ejection of an Auger electron. In general, nuclear states are not
single-particle states (see Fig. 2-30), so that a complicated rearrangement of
nucleons occurs during gamma decay.

The underlying features of electromagnetic emission can be understood
with classical concepts derived from Maxwell's equations. Finer details can be
explained only by using quantum mechanics. The difference in angular momenta
and the relative parities of the nuclear states involved in the transition play a
crucial role in determining the transition probability. We will examine this after
a brief discussion of energetics.

4-4a Energetics of gamma decay.   If the initial excited nucleus has a rest mass
M_0^* and the final state has a rest mass M_0, conservation of energy and momentum
require (see Fig. 4-4)

$$M_0^* c^2 = M_0 c^2 + E_r + T_a \tag{4-43}$$

$$0 = \mathbf{p}_r + \mathbf{p}_a \tag{4-44}$$

---

[1] In analogy with classical electrodynamics, we expect that only charged particles should
radiate. In reality neutron transitions can also produce radiation: first, because the protons in
the nucleus have to be displaced to keep the center of mass fixed; second, because neutrons
as well as protons radiate as a result of their magnetic moments.

FIGURE 4-4 Gamma decay of a nucleus. (*a*) Energy diagram. (*b*) Momentum diagram.

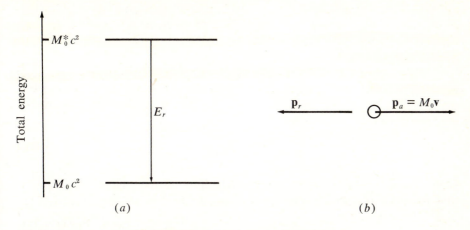

(*a*)   (*b*)

where E_r, $\mathbf{p}_r$ = energy and momentum of gamma ray
    T_a, $\mathbf{p}_a$ = recoil kinetic energy and momentum of final nucleus
The recoil speed of the nucleus is so small that nonrelativistic formulae may be used to compute T_a.

$$T_a = \frac{p_a^2}{2M_0}$$

$$= \frac{p_r^2}{2M_0}$$

$$= \frac{E_r^2}{2M_0c^2} \tag{4-45}$$

using Eqs. (2-1) and (2-3). If, typically, $E_r = 2$ Mev and $A = 50$

$$T_a = \frac{2^2}{2 \times 50 \times 930} \text{ Mev}$$

$$\approx 40 \text{ ev}$$

For most purposes this is negligible.[1] Hence the gamma-ray energy is

$$E_r \approx (M_0^* - M_0)c^2 \tag{4-46}$$

[1] If M_0 represents the ground state mass and the gamma ray of energy E_r is used to excite another nucleus of the same nuclide, the maximum excitation energy which can be reached is $M_0^* - 2T_a$, because the second nucleus will also recoil. Since most bound nuclear levels have widths much less than 1 ev [see Eqs. (4-69)], the state M_0^* cannot be excited unless the recoil energy loss is restored by moving either the source or the target to obtain the appropriate energy by Doppler shift of the gamma-ray energy.

4-4b Decay constant for gamma decay.   Gamma decay of an excited nucleus requires some time, just like the decay of an excited atom. Half-lives of excited atomic states are typically 10^{-8} sec for valence electrons and 10^{-15} sec for hole states formed after electron ejection from an inner electronic shell. Nuclear excited states have half-lives for gamma emission[1] ranging from 10^{-16} sec to longer than 100 years.

These half-lives can be crudely estimated on the basis of semiclassical considerations. It can be shown from Maxwell's equations that an accelerated point charge e radiates electromagnetic radiation at a rate (all quantities in electrostatic units)

$$\frac{dE}{dt} = \frac{2}{3}\frac{e^2 a^2}{c^3} \tag{4-47}$$

where $a = (a_x^2 + a_y^2 + a_z^2)^{\frac{1}{2}}$ is the acceleration of the charge. This formula does not hold for an extended charge distribution because interference effects have to be considered.

To make a simple model of the emission process, let us assume that the radiating charge (electron in atom, proton in nucleus) oscillates with a simple harmonic motion

$$x = x_0 \cos \omega t \tag{4-48}$$

and similarly for y and z. It is reasonable to choose the amplitudes such that

$$x_0^2 + y_0^2 + z_0^2 \approx R^2 \tag{4-49}$$

where R is radius of atom or nucleus. Then

$$a \approx R\omega^2 \cos \omega t \tag{4-50}$$

Substituting into Eq. (4-47), the average energy radiated over many cycles is given by

$$\left(\frac{dE}{dt}\right)_{\text{ave}} \approx \frac{e^2 R^2 \omega^4}{3c^3} \tag{4-51}$$

since $(\cos^2 \omega t)_{\text{ave}} = \frac{1}{2}$.

Even though this expression is derived from classical equations, we must take into account the fact that electromagnetic radiation is radiated in quanta. To make the transition from classical to quantum theory, we assume that each photon is emitted during a mean time interval τ. The average rate of energy emission is then given by

$$\left(\frac{dE}{dt}\right)_{\text{ave}} = \frac{h\nu}{\tau} = \hbar\omega\lambda_f = E_r\lambda_f \tag{4-52}$$

---

[1] If particle decay can occur (see Sec. 2-6), half-lives of nuclear states can be as short as the estimate (2-144) indicates, i.e., 10^{-22} sec.

where we associate τ with the mean life for gamma decay [$\tau = 1/\lambda_\gamma$; see Eq. (4-8)]. (Do not confuse the wavelength λ_r of electromagnetic radiation with the decay constant λ_γ.) Substituting into Eq. (4-51) and noting that $\omega = 2\pi\nu$, we obtain

$$\lambda_\gamma \approx \frac{e^2 R^2 E_r{}^3}{3\hbar^4 c^3} \qquad (4\text{-}53)$$

The mass of the emitting particle does not enter into this expression. Applying it to an atom ($R \approx 10^{-8}$ cm) emitting a 1-ev photon and to a nucleus ($R \approx 5 \times 10^{-13}$ cm) emitting a 1-Mev photon, we find for the atom

$$\lambda_\gamma \approx \frac{(4.80 \times 10^{-10})^2 (10^{-8})^2 (1.60 \times 10^{-12})^3}{3(1.05 \times 10^{-27})^4 (3 \times 10^{10})^3} \approx 10^6 \text{ sec}^{-1}$$

$$t_{\frac{1}{2}} \approx 7 \times 10^{-7} \text{ sec} \qquad \textit{right ballpark for } \Gamma \textit{ decay in atom}$$

and for the nucleus

$$\lambda_\gamma \approx \frac{(4.80 \times 10^{-10})^2 (5 \times 10^{-13})^2 (1.60 \times 10^{-6})^3}{3(1.05 \times 10^{-27})^4 (3 \times 10^{10})^3} \approx 2 \times 10^{15} \text{ sec}^{-1}$$

$$t_{\frac{1}{2}} \approx 3 \times 10^{-16} \text{ sec}$$

Although atomic half-lives are indeed of the order of the calculated magnitude and some nuclear half-lives are as short as estimated, the large range of actual nuclear half-lives for gamma emission shows that some important effects have not been considered in expression (4-53).

4-4c Quantum-mechanical effects. We have previously pointed out that in atomic and nuclear systems the position of a particle is not a meaningful concept, since it cannot be determined without gross disturbance of the system (Heisenberg's uncertainty principle). Only the probability of finding the particle in a particular volume element $dx\, dy\, dz$ can be determined, i.e., $\Psi^* \Psi\, dx\, dy\, dz$. We showed in Sec. 2-2c that for time-independent potentials this quantity is independent of time.

When a system is emitting electromagnetic radiation, it is subjected to a time-dependent potential created by the oscillating electric and magnetic fields in which it is immersed. Therefore, the quantity $\Psi^* \Psi$ is no longer time independent [see Eq. (4-34)]. One finds in a quantum-mechanical formulation of the emission process[1] taking the system from an initial state i to a final state f that the position coordinate x of Eq. (4-48) has to be replaced by a *transition matrix element*

$$\int \Psi_f^* x \Psi_i\, dx\, dy\, dz + \text{c.c.} = \left[\int \psi_f^* x \psi_i\, dx\, dy\, dz \right] e^{i[(E_f - E_i)/\hbar]t} + \text{c.c.}$$

$$= x_{fi} e^{-i\omega t} + \text{c.c.} \qquad (4\text{-}54)$$

[1] Schiff, 1955, sec. 66.

where the quantum condition $\hbar\omega = E_i - E_f$ has been assumed and x_{fi} is an abbreviated notation for the (time-independent) matrix element in brackets. The letters c.c. mean complex conjugate of the preceding expression. The expression for the transition probability is analogous to Eq. (4-53),

$$\lambda_\gamma = \frac{4e^2(|x_{fi}|^2 + |y_{fi}|^2 + |z_{fi}|^2)E_r^3}{3\hbar^4 c^3} \tag{4-55}$$

It is important to discuss certain properties of a typical matrix element

$$x_{fi} = \int \psi_f^* x \psi_i \, dx \, dy \, dz \tag{4-56}$$

First, this matrix element, being a physically observable quantity,[1] cannot depend on the coordinate system used. It must have the same value in a left-handed coordinate system as in a right-handed system; in other words it must be invariant to the parity operation (Sec. 2-2h) $x \to -x$, $y \to -y$, $z \to -z$. Now, if the states i and f have the *same* parity (see Sec. 2-2h), we see from Eq. (4-56) that

$$x_{fi} \to -x_{fi} \tag{4-57}$$

under the parity operation. Therefore x_{fi} must vanish. Since the same reasoning applies to y_{fi} and z_{fi}, we find $\lambda_\gamma = 0$ according to Eq. (4-55). This is our first example of a *selection rule*. A selection rule generally states a condition which is necessary in order for a given process to occur. Here the condition is that the states i and f must have opposite parity in order that the gamma decay described by the matrix element (4-56) takes place (because then $x_{fi} \to x_{fi}$ under the parity operation). If we denote parity by π, the requirement is

$$\pi_i = -\pi_f \tag{4-58}$$

It turns out, even if the parities of i and f are opposite, the matrix element (4-56) still vanishes unless the angular momenta $\mathbf{I}_i$ and $\mathbf{I}_f$ of the states are such that they differ vectorially by unit vector angular momentum (in units of $\hbar$, of course) which means that

$$I_i - I_f = \pm 1 \text{ or } 0 \quad \text{with } I_f = I_i = 0 \text{ forbidden} \tag{4-59}$$

The requirement that the angular momentum vector change during the emission process can be understood classically. One can show that electromagnetic radiation, emitted by an oscillating charge of the type (4-48), carries away angular momentum. This means, if the radiation is absorbed by a large, perfectly absorbing, hollow sphere at whose center the oscillating charge is located, such a sphere would acquire angular momentum. According to the classical calculations, based on Maxwell's equations, angular momentum is lost at a continuous

---

[1] In the simple case (4-55) only the absolute square of the matrix element is observable, but in other situations the matrix element itself can be determined.

rate by the radiating system. Quantum calculations show, though, that the angular momentum change must occur in a discrete step. [The same difference occurs in the energy emission process—see Eq. (4-52).]

4-4d Classification of gamma decays.   It would seem from the preceding discussion that no gamma decay can take place between nuclear levels unless they fulfill the very restrictive conditions (4-58) and (4-59). But the entire derivation assumed that the electromagnetic radiation was generated by a *point* charge in motion. In reality, the nucleus is an *extended* charge distribution in which currents flow, generated by the orbital as well as spin motions of the nucleons. The electric and magnetic fields produced in a transition are therefore much more complicated than implied in Eq. (4-47).

In a classical calculation one proceeds as follows. The actual charge-current distribution is expanded in *multipole moments* each of which has given dimensionality. For a static distribution of charges e_i located at (x_i, y_i, z_i) the multipole moments have a dimensionality $\sum_i e_i x_i^L$ for a moment of order L.

For example, if $L = 0$, $\sum_i e_i$ is simply the total charge. If $L = 1$, $\sum_i e_i x_i$ is a component of the electric dipole moment of the system. If the particles carry magnetic moments as well as charges one can also expand the distribution of magnetization in magnetic multipole moments.

When the charges oscillate, each multipole moment emits a characteristic electric and magnetic field pattern (except the moment with $L = 0$). The emitted fields can be grouped first by the order of the emitting moment and second by the effect of the parity operation. Conventionally, we speak of *electric* and *magnetic multipole radiation*, although there is generally no simple correspondence with the static multipole fields. For example, electric dipole radiation, which is assumed by Eq. (4-48), emits an electric field which changes sign under the parity operation, while magnetic dipole radiation created by an oscillating current loop, for example, does not change sign (see Fig. 4-5).[1]

Although in a classical radiation problem the parity property is of no particular importance, it is very important for the gamma decay between nuclear states. Indeed, if the particular multipole radiation between the initial state i and the final state f produces a parity change π_r, conservation of parity would require

$$\pi_i = \pi_f \pi_r \tag{4-60}$$

Experimentally it has been found that in electromagnetic decays this selection rule is obeyed to a very high order of precision.

In quantum mechanical calculations[2] each multipole moment of order L_r is found to produce radiation which carries off an angular momentum $\mathbf{L}_r \hbar$,

---

[1] Fig. 4-5 is only schematic. For a full treatment see De Benedetti, 1964, sec. 1.15.
[2] Blatt and Weisskopf, 1952, chap. 12, sec. 2.

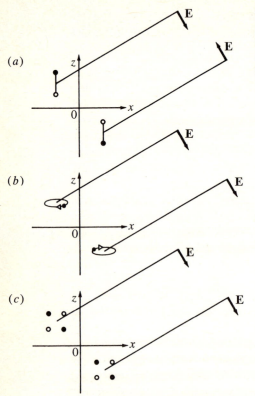

(a)

(b)

(c)

FIGURE 4-5 Effect of the parity operation on the electric field **E** radiated according to classical theory by (a) an electric dipole, (b) a magnetic dipole, (c) an electric quadrupole. (By permission from Burcham, 1963.)

so that conservation of angular momentum requires

$$\mathbf{I}_i = \mathbf{I}_f + \mathbf{L}_r \qquad (4\text{-}61)$$

The parity change π_r is directly related to L_r, and

For electric multipole radiation, $\pi_r = (-1)^{L_r}$

For magnetic multipole radiation, $\pi_r = -(-1)^{L_r}$ $\qquad (4\text{-}62)$

Table 4-1 gives the classification of a few common types of radiation. We see that the radiation considered in Sec. 4-4c is $E1$ radiation.

Although this classification may seem complicated, a great simplification is brought about by the fact that in practice in a given transition usually one and at most two multipole radiations are of importance. This occurs because in the expression for the gamma transition probability

$$\lambda_\gamma = \lambda_\gamma(E1) + \lambda_\gamma(M1) + \lambda_\gamma(E2) + \cdots \qquad (4\text{-}63)$$

a certain set of terms is eliminated by the selection rules (4-60) and (4-61), and for the remainder the decay constant of the multipole of lowest order usually

TABLE 4-1   Classification of gamma radiation

Name	Abbreviation	L_r	π_r
Electric dipole	E1	1	−1
Magnetic dipole	M1	1	+1
Electric quadrupole	E2	2	+1
Magnetic quadrupole	M2	2	−1
Electric octupole	E3	3	−1

TABLE 4-2   Examples of gamma decays

Initial state†	Final state	Predominant decay mode
2^+	0^+	$E2$‡
1^+	0^+	$M1$‡
$\frac{1}{2}^-$	$\frac{1}{2}^+$	$E1$‡
2^+	2^+	$M1$
$\frac{9}{2}^+$	$\frac{1}{2}^-$	$M4$
0^+	0^+	no gamma decay

† The total angular momentum and parity of each state is given.
‡ Only possible decay mode for this transition.

exceeds that of all the other multipoles by a factor of at least 10^2 to 10^4. In Table 4-2 the selection rules (4-61) and (4-62) have been applied to a few specific examples.

Theoretical estimates of the decay constants are only approximate, because the nuclear wave functions which enter into transition matrix elements such as (4-56) are only approximately known. For a single-proton transition in which the final state is an s state, Weisskopf[1] has estimated that for a nucleus of radius R

$$\lambda_\gamma(EL_r) \approx S \frac{e^2}{\hbar \lambdabar_r} \left(\frac{R}{\lambdabar_r} \right)^{2L_r} \tag{4-64}$$

and

$$\lambda_\gamma(ML_r) \approx 10 \left(\frac{\hbar}{M_p c R} \right)^2 \lambda_\gamma(EL_r) \tag{4-65}$$

where S is a statistical factor and M_p is the proton mass.

$$S = \frac{2(L_r + 1)}{L_r[1 \times 3 \times 5 \cdots (2L_r + 1)]^2} \left(\frac{3}{L_r + 3} \right)^2 \tag{4-66}$$

[1] Blatt and Weisskopf, 1952, p. 627.

λ_r is the wavelength of the electromagnetic radiation divided by 2π. From Eq. (2-2)

$$\lambda_r(\text{in F}) = \frac{197}{E_r\ (\text{in Mev})} \qquad (4\text{-}67)$$

Typical values of R/λ_r are therefore around $1/40$ ($R = 5$ F, $E_r = 1$ Mev) so that, even without the factor S, multipoles differing by unity in their order differ by 10^{-3} in decay rate. In addition, S decreases by a factor of about 10^{-2} for each order.[1] Assuming $R = 1.2A^{\frac{1}{3}}$ F,

$$\lambda_\gamma(ML_r) \approx 0.3A^{-\frac{2}{3}}\lambda_\gamma(EL_r) \qquad (4\text{-}68)$$

so that magnetic multipole radiation is less probable than electric multipole radiation of the same order. Note, though, that the two can never occur together because their parities differ according to Eq. (4-62).

Figure 4-6 plots the results of Eqs. (4-64) and (4-65) for a nucleus with $A = 55$. Actually for $L_r = 1$, Eq. (4-64) gives practically the same result as the semiclassical estimate (4-53). We can also use Eqs. (4-64) and (4-65) to estimate

[1] R. D. Evans, 1955, p. 214.

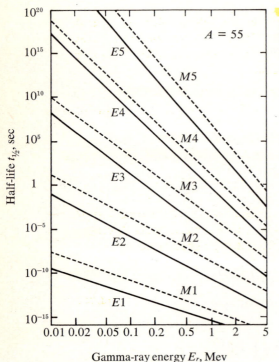

FIGURE 4-6 Half-life for gamma-ray multipole emission according to the Weisskopf estimate (4-64) and (4-65) for a nucleus with $A = 55$. (By permission from A. H. Wapstra, G. J. Nijgh, and R. Van Lieshout, "Nuclear Spectroscopy Tables," North Holland Publishing Company, Amsterdam, 1959, as adapted by Burcham, 1963.)

gamma-decay widths (Sec. 4-3). With Γ_γ in ev and E_r in Mev,

$$\Gamma_\gamma(E1) = 0.068 E_r^3 A^{\frac{2}{3}}$$
$$\Gamma_\gamma(M1) = 0.021 E_r^3$$
$$\Gamma_\gamma(E2) = 4.9 \times 10^{-8} E_r^5 A^{\frac{4}{3}} \qquad (4\text{-}69)$$
$$\Gamma_\gamma(M2) = 1.5 \times 10^{-8} E_r^5 A^{\frac{2}{3}}$$
$$\Gamma_\gamma(E3) = 2.3 \times 10^{-14} E_r^7 A^2$$

Figures 4-7 and 4-8 compare some experimental mean lives τ with the Weisskopf estimates (4-64) and (4-65). In most cases the mean life is longer than estimated.[1] One exceptional group is formed by the $E2$ transitions, which are appreciably more rapid than predicted in regions of nucleon numbers between closed shells. This can be understood on the basis of the collective model (Sec. 2-5d) because oscillating and rotating nuclei imply a coherent motion of several nucleons. In the decay constant (4-64), we must, therefore, insert $(ne)^2$ instead of e^2, where n is the effective number of nucleons moving coherently. There are also other ways to express the coherence effect; in particular the $E2$ decay constant of a deformed nucleus can be related to its static electric quadrupole moment.

4-4e Internal conversion. The electric and magnetic fields, which are created for a very short time as one or more nucleons in the nucleus rearrange themselves during a transition from an initial to a final state, can give rise to another transition process, known as internal conversion. The nuclear energy $E_i - E_f$ is transferred *directly* to an atomic electron, which is ejected with a kinetic energy

$$T_e = E_i - E_f - E_B \qquad (4\text{-}70)$$

where E_B is the binding energy of the electron in the atomic shell from which it has been ejected. (The recoil energy of the atom has been neglected in this equation.) Although Eq. (4-70) resembles Eq. (3-49) of the external photoelectric effect in form, internal conversion is not an internal photoelectric effect, but an additional process by which a nucleus can release excitation energy, besides gamma emission. This comes about because internal conversion is produced by the (time varying) coulomb field of the nucleus, which has a *radial* direction. Gamma emission is caused by *transverse* electric and magnetic fields. Different field components enter in the two processes and hence the processes are independent. The total decay constant for deexcitation of the state is, therefore,

$$\lambda_{\text{tot}} = \lambda_\gamma + \lambda_e \qquad (4\text{-}71)$$

where λ_e is the decay probability by internal conversion and λ_γ has been estimated

---

[1] This is expressed by the concept of *hindrance factor*, shown in Fig. 4-7 for $E1$ and $M1$ transitions. The hindrance factor is the ratio $\tau(\text{experimental})/\tau(\text{Weisskopf})$, where τ is the mean life for gamma decay ($= 1/\lambda_\gamma$).

FIGURE 4-7 Comparison of experimental mean lives for gamma decay with Weisskopf estimates [Eqs. (4-64) and (4-65)] versus gamma-ray energy. The ratio of the lifetimes is called hindrance factor. Top curve for $E1$ transitions, bottom curve for $M1$ transitions. (Adapted from N. B. Gove, Beta and Gamma Transition Probabilities, in N. B. Gove and R. L. Robinson, (eds.), "Nuclear Spin and Parity Assignments," Academic Press Inc., New York, 1966.)

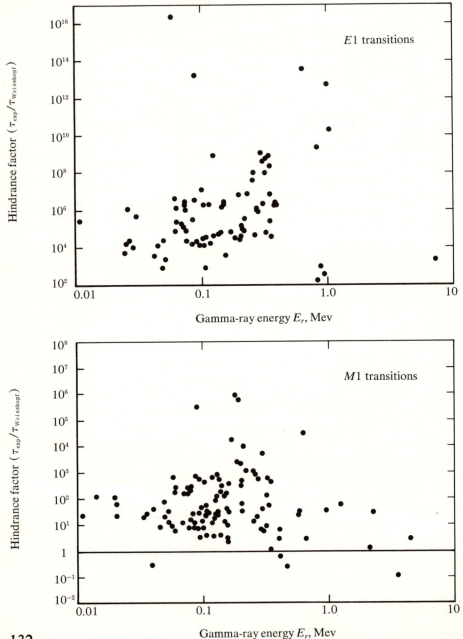

in the preceding sections. Depending on the origin of the ejected electrons, λ_e can be decomposed into the decay probability for $K, L, M \cdots$ emission

$$\lambda_e = \lambda_K + \lambda_L + \lambda_M + \cdots \tag{4-72}$$

Experimentally, these processes can be distinguished by the different energies of the emitted electrons.

Although the calculations of the absolute decay constant λ_e is beset by difficulties similar to the calculation of the gamma decay constant λ_γ, the ratio

$$\frac{\lambda_e}{\lambda_\gamma} = \alpha \tag{4-73}$$

FIGURE 4-8 Reduced gamma-ray mean life for electric quadrupole transitions against neutron number. The Weisskopf estimate (4-64) corresponds to the horizontal line shown. (By permission from W. M. Currie, "*E2* Mean-Lives in Even-Even Nuclei," *Nuclear Data Sheets*, vol. 5, set 2, 1962.)

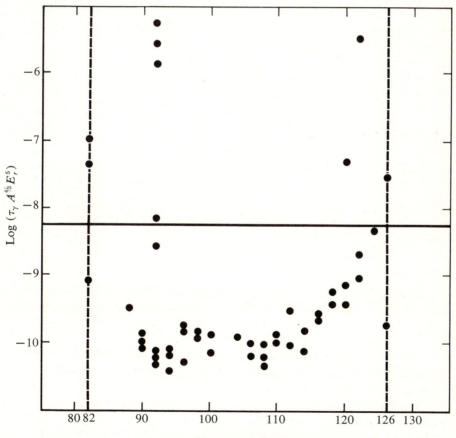

which is called conversion coefficient, can be computed to within a few percent. The usefulness of internal conversion for nuclear structure studies lies in the fact that for a given energy difference $E_i - E_f$ and for a given atomic number Z of the decaying nucleus, the conversion coefficient depends sensitively on the type and multipolarity of the accompanying electromagnetic transition. It is an increasing function of L_r and Z and a decreasing function of $E_i - E_f$. Typical values of a K conversion coefficient $\alpha_K = \lambda_K / \lambda_\gamma$ are shown in Fig. 4-9.

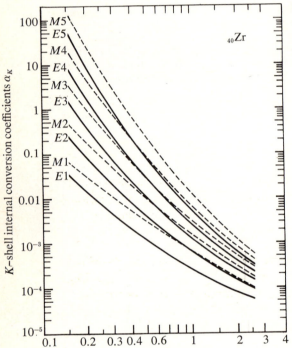

FIGURE 4-9 *K*-shell internal conversion coefficient for $Z = 40$ plotted against transition energy. (By permission from Evans, 1955.)

Although gamma decay between two nuclear states both with spins and parity 0^+ is absolutely forbidden, decay by internal conversion can take place. Transitions of the kind $0^+ \rightarrow 0^-$ can occur only by two-photon emission, but have not yet been detected.

Internal conversion is always accompanied by a secondary process because the atom is left in an excited state of energy E_B. The energy is released by x rays or Auger electrons (see Sec. 3-4c).

For transition energies $E_i - E_f \geq 2m_0 c^2$, where m_0 is the electron rest mass, decay can also take place by creation of an electron-positron pair. As in internal conversion, this internal pair formation is an additional decay mode consisting in a direct transfer of the decay energy to the *virtual* negative-energy electrons (see Fig. 3-23). The process has a probability of approximately 10^{-3} compared to gamma decay. In the case of $0^+ \rightarrow 0^+$ transitions, it can predominate over internal conversion if the decay energy is high enough.

4-4f Nuclear structure information from gamma decay.   A study of gamma decay reinforces the usefulness of nuclear models. In particular, the single-particle shell model provides the guiding concepts in an understanding of the decay probability. Since the decay is determined by the spatial overlap of the initial and final wave functions [see, for example, expression (4-56)], fine details of the wave functions become apparent. Presently, they are not all clear and form the basis of further research. In a few cases of nuclei removed by one nucleon from doubly magic numbers, it has been possible to reproduce experimental decay constants. Also, for permanently deformed nuclei, the decay probability between rotational states (Sec. 2-5d) is well understood.

4-5 ALPHA DECAY

Alpha radioactivity has been investigated for a long time because the naturally radioactive substances which led to the discovery of radioactivity (Becquerel, 1896) were found to be alpha emitters (Curie, Rutherford). From the point of view of nuclear structure, alpha decay represents particle decay of a virtual nuclear level (Fig. 2-29) and can serve as a prototype for this phenomenon. We will encounter this type of decay again in our study of nuclear reactions (Sec. 5-5).

As we will elaborate below, most nuclides with $A > 150$ are unstable against alpha decay. For the lighter nuclides, alpha decay is very improbable. The decay constant decreases exponentially with decreasing decay energy and, close to $A = 150$, the decay energy is practically zero (Fig. 4-11). The nuclides near $N = 82$ are exceptional because shell effects provide additional decay energy.

In the experimental information on alpha decay, several systematic trends are apparent. First, the dependence of the decay energy on A (or Z or N) is regular except near magic numbers. The trend is in accord with the semiempirical mass formula[1] (Sec. 2-4). Second, for nuclides with a given Z, the half-life is a smooth function of the decay energy, especially for even-even nuclei. This relationship reflects the decay mechanism. Third, the energy spectra of alpha particles give information about the level schemes of the parent or daughter nuclei.

4-5a Energetics of alpha decay.   If the parent nucleus P has a nuclear mass M'_P and the daughter nucleus D has a nuclear mass M'_D, conservation of energy and

---

[1] See Probs. 4-14 and 4-15.

momentum require (see Fig. 4-10)

$$M'_P c^2 = M'_D c^2 + T_D + M'_\alpha c^2 + T_\alpha \qquad (4\text{-}74)$$

$$0 = \mathbf{p}_D + \mathbf{p}_\alpha \qquad (4\text{-}75)$$

where M'_α = nuclear mass of alpha particle
$T_\alpha, \mathbf{p}_\alpha$ = kinetic energy and momentum of alpha particle
$T_D, \mathbf{p}_D$ = recoil kinetic energy and momentum of daughter nucleus

FIGURE 4-10 Alpha decay of a nucleus. (*a*) Energy diagram. (*b*) Momentum diagram.

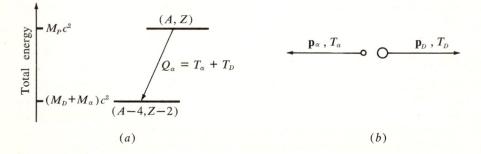

(*a*)   (*b*)

Since it is usual to tabulate atomic rather than nuclear masses, it is convenient to add the masses of the atomic electrons, and their binding energies, to both sides of Eq. (4-74). The electron masses balance exactly and their binding energies balance within a few ev, so that we can write Eq. (4-74) in terms of atomic masses

$$M_P c^2 = (M_D + M_\alpha)c^2 + T_D + T_\alpha \qquad (4\text{-}76)$$

The *decay energy Q_α* is defined as the sum of the resultant kinetic energies

$$Q_\alpha = T_D + T_\alpha \qquad (4\text{-}77)$$

From Eq. (4-76), Q_α is also equal to the difference between the initial and final masses

$$Q_\alpha = [M_P - (M_D + M_\alpha)]c^2 \qquad (4\text{-}78)$$

It is typical of this decay, as well as of nuclear reactions, that Q values can be determined either by particle spectroscopy, i.e., kinetic energy measurements, or by mass spectroscopy. The identity of the measured Q values then demonstrates mass-energy equivalence and conservation, on which Eq. (4-76) is based.

The kinetic energies T_D and T_α are small enough so that nonrelativistic expressions may be used to evaluate them.

$$T_D = \frac{p_D^2}{2M_D}$$

$$= \frac{p_\alpha^2}{2M_D}$$

$$= \frac{M_\alpha}{M_D} T_\alpha \qquad\qquad (4\text{-}79)$$

The recoil kinetic energy T_D is not negligible, as it is in gamma decay. Substituting Eq. (4-79) in Eq. (4-77)

$$Q_\alpha = \frac{M_D + M_\alpha}{M_D} T_\alpha$$

$$\approx \frac{A}{A-4} T_\alpha \qquad\qquad (4\text{-}80)$$

where A is the mass number of the parent. The alpha-particle kinetic energy T_α is always less than the decay energy Q_α.

It is clear from the above considerations that alpha decay cannot take place unless Q_α is positive. Referring to Eq. (2-119) and the corresponding definition of alpha separation energy S_α, we see from Eq. (4-78) that

$$Q_\alpha = -S_\alpha \qquad\qquad (4\text{-}81)$$

Hence, Q_α can be related to the total binding energies of the nuclei by Eq. (2-121)

$$Q_\alpha = B_{\text{tot}(D)}(A-4, Z-2) + B_{\text{tot}(\alpha)}(4,2) - B_{\text{tot}(P)}(A,Z) \qquad (4\text{-}82)$$

Substitution of the semiempirical binding-energy equation (2-127) gives the regions of alpha instability indicated in Fig. 4-11. For the stable nuclei, we find $Q_\alpha > 0$ for $A > 150$. The curves shown do not include shell effects. From Eq. (4-82) we can see, though, that whenever the daughter nucleus $(A - 4, Z - 2)$ is magic, i.e., has a large binding energy, the alpha-decay energy is particularly high. Conversely, whenever the parent nucleus is magic, the decay energy is particularly low. This is demonstrated in a striking fashion in Fig. 4-12. For $N_D = 126$, Q_α is large; for $N_P = 126$, it is small. Similarly for $Z_D = 82$, Q_α is large; for $Z_P = 82$, Q_α is negative and no alpha decay exists.[1] Also, in the rare-earth region, alpha decay is found for nuclides with $N_D \geq 82$ because of these shell effects.

Returning once more to Fig 4-11, we should note that the various curves refer to the *ground* states of the nuclei involved. Sufficiently highly *excited* states of nuclei can emit alpha particles in any region of A and Z, since for

---

[1] Negative Q_α values are not shown on Fig. 4-12 which refers only to observed alpha-decay energies.

FIGURE 4-11 Stability limits predicted by the semiempirical mass formula. Stability limits for various values of Q_α (in Mev) are shown. The region of beta-stable even-even nuclei is indicated by cross hatching. Stability limits for n, p, and d emission (by odd N, odd-Z, and odd-odd nuclei, respectively) are also given. (By permission from G. C. Hanna, Alpha Radioactivity, in E. Segrè, (ed.), "Experimental Nuclear Physics," vol. 3, John Wiley & Sons, Inc., New York, 1959.)

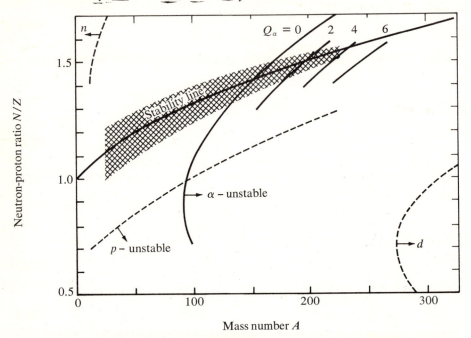

sufficiently large values of M_P the right-hand side of Eq. (4-78) can always be made positive. Figure 4-11 also shows the regions for neutron, proton, and deuteron instability predicted by the semiempirical mass formula. We see that ground states of nuclei lying close to the stability line (Sec. 2-4c) are stable against these decay modes. The reason alpha decay occurs has to do with the exceptionally large value of $B_{tot(\alpha)}$ ($= 28.3$ Mev), which allows Q_α in Eq. (4-82) to be positive for a certain region of Z and N near the stability line.

4-5b Decay constant for alpha decay. The first recognition of systematic trends in the decay constant for alpha decay was made by Geiger and Nuttall (1911). They found a linear relationship between the logarithm of the decay constant and the logarithm of the range of alpha particles from a given naturally radioactive decay chain. It has since been found that this relationship is based on decay-energy and lifetime systematics and is valid only over a limited range of nuclides. Recent experiments show that for ground state decays between even-even nuclides the following relation is valid, which will be derived below,

$$\log t_{\frac{1}{2}} = a + \frac{b}{\sqrt{Q_\alpha}} \tag{4-83}$$

a and b are functions of Z. If Q_α is expressed in Mev and $t_{\frac{1}{2}}$ in seconds one finds[1]

$$a \approx -1.61 Z_D^{\frac{2}{3}} - 21.4$$
$$b \approx 1.61 Z_D \tag{4-84}$$

[1] Segrè, 1964, p. 278.

FIGURE 4-12   Alpha-decay energy versus neutron number of the parent for various proton numbers. (Adapted from Segrè, 1964.)

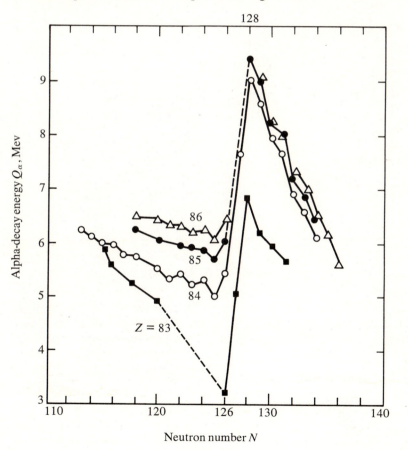

The available data are plotted in Fig. 4-13 according to Eq. (4-83). For decays to excited states or for decays between odd-A (or odd-odd) nuclei, the half-lives are usually longer than for neighboring even-even nuclei with the same decay energy. The multiplicative factor by which such a half-life is longer is called *hindrance factor*. It is, of course, the task of any theory of alpha decay to explain relation (4-83) as well as the hindrance factors.

Physicists who tried to account for alpha emission before the discovery of

FIGURE 4-13 Half-life decay-energy systematics for heavy even-even nuclei. The data are plotted in accordance with Eq. (4-83), i.e., the ordinate is on a logarithmic scale and the abscissa is on a scale varying as $\sqrt{Q_\alpha}$. Points corresponding to the same Z_P are connected by straight lines. [Adapted from C. J. Gallagher and J. O. Rasmusson, *J. Inorg. Nucl. Chem.* **3**: 333 (1957).]

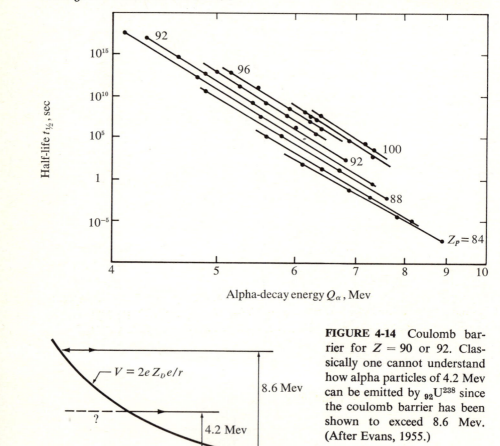

FIGURE 4-14 Coulomb barrier for $Z = 90$ or 92. Classically one cannot understand how alpha particles of 4.2 Mev can be emitted by $_{92}U^{238}$ since the coulomb barrier has been shown to exceed 8.6 Mev. (After Evans, 1955.)

quantum mechanics faced the following **dilemma**. It was known, for example, that no breakdown of Rutherford's alpha-particle scattering law occurred with 8.6-Mev alpha particles on $_{92}U^{238}$ (see Sec. 1-2b). Hence, up to a certain distance from the $_{92}U^{238}$ nucleus, the potential encountered by the alpha particles is pure coulomb, as shown on Fig. 4-14. Yet, $_{92}U^{238}$ emits 4.2-Mev particles, forming $_{90}Th^{234}$. Since the coulomb potential does not change much between U and Th, how could the alpha particle ever get out of the nucleus, i.e., *over* the potential barrier, which must exceed 8.6 Mev?

FIGURE 4-15 Mechanism of alpha decay according to theory of Gamow and of Gurney and Condon. The alpha particle exists inside the potential well formed by nuclear and coulomb forces. The amplitude of its wave function is large inside the well, but there is a small probability of leakage through the potential barrier. (After Burcham, 1963.)

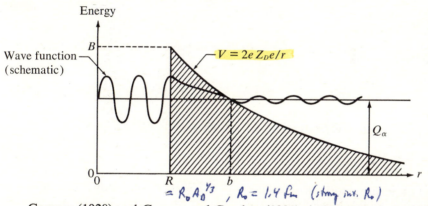

Gamow (1928) and Gurney and Condon (1928), independently, were able to provide the answer to this question by a quantum-mechanical calculation, which we will now discuss. These authors assumed that the alpha particle exists within the nucleus, confined by a nuclear potential approximately as shown in Fig. 4-15. The potential within the nucleus is assumed to be zero to simulate the coulomb effect inside the nucleus. The exact depth of the potential within the nucleus does not affect the final result materially. Although it is believed, today, that the alpha particles do not preexist inside the nucleus with high probability, but are formed in the surface region, the theory gives a good account of relation (4-83).

From a semiclassical point of view, the probability of decay per unit time λ_α is equal to the number of collisions per second which the alpha particle makes with the wall of the confining potential well multiplied by the probability P that the particle penetrates the potential barrier. Within a small numerical factor

$$\lambda_\alpha \approx \frac{v_{\text{in}}}{R} P \qquad (4\text{-}85)$$

where v_{in} is the alpha-particle speed inside the nucleus. Various approximations can be made for P as long as a semiclassical approach is used, the simplest being Eq. (2-106):

$$P \approx e^{-\gamma} \tag{4-86}$$

where γ is given by Eq. (2-108). Calling ze the charge of the alpha particle, we keep the formulae general for later application,

$$\gamma = \frac{2}{\hbar} \int_R^b \left[2M_0 \left(\frac{zZ_De^2}{r} - Q_\alpha \right) \right]^{\frac{1}{2}} dr \tag{4-87}$$

The distance b is indicated in Fig. 4-15. Because of recoil effects, the reduced mass M_0 of the alpha particle appears in this equation:

$$M_0 = \frac{M_\alpha M_D}{M_\alpha + M_D} \tag{4-88}$$

(see Sec. 2-2e). The integral in Eq. (4-87) can be evaluated in a straightforward manner giving

$$\gamma = \frac{4zZ_De^2}{\hbar v} \left[(\cos^{-1} \sqrt{y}) - \sqrt{y}(1 - y)^{\frac{1}{2}} \right] \tag{4-89}$$

where v = relative velocity of alpha particle and daughter nucleus

$$y = R/b = Q_\alpha/B \tag{4-90}$$

The last identity follows from Eqs. (4-91) and (4-92) below. The coulomb barrier height B is given by (see Fig. 4-15)

$$B = \frac{zZ_De^2}{R} \tag{4-91}$$

and we note that

$$Q_\alpha = \frac{1}{2} M_0 v^2 = \frac{zZ_De^2}{b} \tag{4-92}$$

by definition of the turning point b (see Figs. 2-6 and 4-15).

For *thick barriers*, i.e., $b \gg R$ or $Q_\alpha \ll B$, we can expand the bracket[1] in Eq. (4-89)

$$(\cos^{-1} \sqrt{y}) - \sqrt{y}(1 - y)^{\frac{1}{2}} \approx \tfrac{1}{2}\pi - 2\sqrt{y} \tag{4-93}$$

to obtain

$$\gamma \approx \frac{2\pi zZ_De^2}{\hbar v} - \frac{4}{\hbar}(2zZ_De^2M_0R)^{\frac{1}{2}} \tag{4-94}$$

[1] Evans, 1955, p. 876.

Summarizing Eqs. (4-85), (4-86), and (4-94) for alpha decay through a thick barrier

$$\lambda_\alpha \approx \frac{v_{in}}{R} \exp\left[-\frac{4\pi Z_D e^2}{\hbar v} + \frac{8}{\hbar}(Z_D e^2 M_0 R)^{\frac{1}{2}} \right] \qquad (4\text{-}95)$$

Substituting for v from Eq. (4-92), we see that the form of Eq. (4-83) and the Z dependence (4-84) are reproduced. An increase in Z_D will thicken the barrier, hence decrease λ_α. An increase in R will decrease the barrier thickness, hence increase λ_α.

To appreciate the order of magnitude of the terms in Eq. (4-95), we will evaluate it for the 4.2-Mev alpha particles from $_{92}U^{238}$, ignoring recoil effects. From Eq. (4-92)

$$v \approx \left(\frac{2 \times 4.2 \times 1.60 \times 10^{-6}}{4 \times 1.65 \times 10^{-24}} \right)^{\frac{1}{2}}$$

$$\approx 1.43 \times 10^9 \text{ cm/sec}$$

$$R = 1.4(234)^{\frac{1}{3}} \, 10^{-13}$$

$$= 8.6 \times 10^{-13} \text{ cm}$$

Since for the potential assumed (Fig. 4-15), the speed v_{in} of the particle inside the nucleus is the same as the speed v far from the nucleus

$$\frac{v_{in}}{R} \approx 1.7 \times 10^{21} \text{ sec}^{-1} \qquad (4\text{-}96)$$

The first term of the exponent in Eq. (4-95) is

$$\frac{-4\pi Z_D e^2}{\hbar v} = \frac{-4\pi \times 90 \times (4.80 \times 10^{-10})^2}{1.05 \times 10^{-27} \times 1.43 \times 10^9}$$

$$= -173 \qquad (4\text{-}97)$$

The second term of the exponent is

$$\frac{8}{\hbar}(Z_D e^2 M_0 R)^{\frac{1}{2}} = \frac{8}{1.05 \times 10^{-27}} [90 \times (4.80 \times 10^{-10})^2 \times 4 \times 1.65 \times 10^{-24}$$

$$\times 8.6 \times 10^{-13}]^{\frac{1}{2}}$$

$$= 83 \qquad (4\text{-}98)$$

Hence

$$P = e^{-90} \approx 10^{-39} \qquad (4\text{-}99)$$

indicating the extremely small transmission probability typical of alpha decay. Combining this with Eq. (4-96)

$$\lambda_\alpha \approx 1.7 \times 10^{-18} \text{ sec}^{-1}$$

$$t_{\frac{1}{2}} \approx 4.1 \times 10^{17} \text{ sec} = 1.3 \times 10^{10} \text{ years}$$

The experimental half-life is 0.45×10^{10} years, a remarkable agreement considering the simplifications made.

From Eq. (4-91) we can evaluate

$$B = \frac{2 \times 90 \times (4.80 \times 10^{-10})^2}{8.6 \times 10^{-13}} \text{ ergs}$$

$$= 30 \text{ Mev} \tag{4-100}$$

and from Eq. (4-90) $b = \dfrac{RB}{Q_\alpha} = \dfrac{8.6 \times 10^{-13} \times 30}{4.2}$

$$= 61 \times 10^{-13} \text{ cm} \tag{4-101}$$

This is indeed a thick barrier and justifies the approximation (4-93).

A comparison of Eq. (4-95) with the experimental material (Fig. 4-13) allows a determination of R, because λ_α is very sensitive to R. We see from the evaluation (4-98) that a 2-percent change in R changes λ_α roughly by a factor of 2. The available data are consistent with the relation $R = R_0 A^{\frac{1}{3}}$ and require R_0 to lie between 1.4 and 1.5 F.

FIGURE 4-16   Effect of angular momentum change in alpha decay.   (a) Classical interpretation.   (b) Modification of the effective barrier thickness.

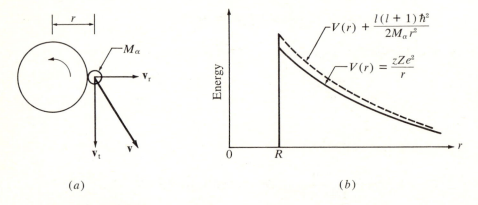

(a)   (b)

4-5c Hindrance factors.   The theory as presented above applies only to ground-state decays between even-even nuclei because then no angular momentum is carried off by the alpha particle. If the decay takes place from an excited state of the parent or to an excited state of the daughter (see Sec. 4-5d), an angular momentum change will generally be involved (see Fig. 2-28 for a typical level scheme). This affects the decay constant.

Even classically we can imagine that the alpha particle leaves the nucleus in such a manner that the daughter nucleus acquires angular momentum. If, as indicated in Fig. 4-16a, the alpha particle leaves with a velocity **v**, which has a tangential component v_t, conservation of angular momentum[1] requires that the daughter nucleus receive an angular momentum equal to

$$L = M_\alpha v_t r \tag{4-102}$$

Also, the kinetic energy of the alpha particle can be divided into a radial part

$$\tfrac{1}{2} M_\alpha v_r^2 \tag{4-103}$$

and a tangential part, which can be written

$$\tfrac{1}{2} M_\alpha v_t^2 = \frac{L^2}{2M_\alpha r^2} \tag{4-104}$$

This follows a derivation given already in Sec. 2-2d. Neglecting recoil effects, conservation of energy requires

$$\tfrac{1}{2} M_\alpha v_r^2 + \frac{L^2}{2M_\alpha r^2} + V(r) = E \tag{4-105}$$

The second term on the left can be considered as a *centrifugal potential energy* and can be combined with the potential energy $V(r)$. Angular momentum change in alpha decay, therefore, increases the effective thickness of the barrier (Fig. 4-16b) and increases the half-life of the decay. The increase turns out to depend on the ratio[2]

$$\sigma = \frac{\text{centrifugal barrier height}}{\text{coulomb barrier height}}$$

$$= \frac{l(l+1)\hbar^2}{2M_\alpha zZe^2R}$$

$$\approx 0.002 l(l+1) \qquad \text{for } Z \approx 90 \qquad R \approx 10^{-12} \text{ cm} \tag{4-106}$$

and consists in multiplying the second term in Eq. (4-94) or in the exponent of Eq. (4-95) by $1 - \tfrac{1}{2}\sigma$. In the example at the end of Sec. 4-5b, the hindrance factor so obtained for $l = 2$ would be equal to [see Eq. (4-98)]

$$\exp (83 \times \tfrac{1}{2} \times 0.002 \times 2 \times 3) \approx 1.6$$

which is, of course, very small compared to the effect of Q_α or R.

In odd-A or odd-odd nuclei the shell model predicts a reduced probability of finding an alpha-particle configuration within a nucleus. Alpha-decay from such nuclei is, therefore, hindered with respect to even-even nuclei, even if no orbital angular momentum is carried off by the alpha particle.

---

[1] Conservation of angular momentum takes place about the center of mass. For simplicity, we assume in Eqs. (4-102) to (4-106) that the center of mass is practically at the center of the daughter nucleus.

[2] R. D. Evans, 1955, p. 877.

Alpha decay from oriented nuclei has shown that nuclei far from the closed shell configurations are indeed deformed as suggested by the collective model (Sec. 2-5d). In these nuclei, the nuclear potential takes on the ellipsoidal shape of the mass distribution. Where the barrier is thinnest, preferential decay probability is expected. This has been found by aligning the nuclei and observing the angular distribution of the alpha particles with respect to the direction of nuclear alignment.

4-5d Alpha-particle spectra.   In many alpha decays the daughter nuclei can be left in several excited states, giving rise to *fine structure* in the alpha spectra. A typical decay is shown in Fig. 4-17a. The differences in the *branching ratios*[1] are mainly due to the decay-energy differences, although there is some influence from the angular momentum carried off by the alpha particle.

Alpha decay can also take place from excited states of the parent, as shown in Fig. 4-17b for Po^{212}, producing *long-range* alpha particles. In that case, the intensity of each alpha branch depends (1) on the branching *to* the level, and (2) on the competition between alpha and gamma decay.

Both types of alpha spectra are very useful in the level scheme investigations of nuclei, especially if they are combined with gamma-decay studies. In this way one has been able to delineate the applicability of the shell model and the collective model to various regions of the periodic table above $A = 150$.

4-6   BETA DECAY

Beta decay is the most common type of radioactive decay because all nuclides not lying in the valley of stability (see Fig. 4-11 or Fig. 2-15) are susceptible to beta activity. The process consists in the emission of an electron directly from a nucleus. Both positive and negative electrons can be emitted, in some cases from the same nuclide. Rutherford and Soddy (1903) demonstrated by chemical means that the atomic number of a nuclide *increases* by unity during *negative* beta decay. It was later shown that the atomic number decreases by unity during positron radioactivity (discovered by Curie and Joliot, 1934). Initial investigation of beta radioactivity confused conversion electrons with the electrons emitted from the nucleus. Chadwick (1914) showed, though, that whereas the former are monoenergetic the latter have a continuous energy distribution for a given nuclide.[2]

4-6a Neutrino hypothesis.   The continuous energy distribution of electrons (or positrons) in beta decay proved to be a great puzzle, although the maximum energy of the distribution corresponds exactly to that expected from the mass difference of the parent and the daughter. Neglecting the recoil of the daughter

---

[1] The branching ratio is the fractional decay probability, usually expressed in percent, from a given initial state of a system to one of several possible final states of the system.

[2] A single nucleus emits only one electron with a given kinetic energy.

FIGURE 4-17 Typical alpha-decay schemes. (a) Decay to excited states of a daughter nucleus. Spins and parities of the states are indicated. (b) Decay from excited states of a parent nucleus (Po²¹²). The numbers adjacent to each transition give the intensities relative to 100 Po²¹² ground-state transitions. (Adapted from Segrè, 1964, and from K. Way, N. B. Gove, C. L. McGinnis, R. Nakasima, Energy Levels of Nuclei, $A = 21$ to $A = 212$, and J. Scheer, Energy Levels of the Heavy Nuclei, $A = 213$ to $A = 257$, in A. M. and K. H. Hellwege, Editors, Landolt-Bornstein, group 1, vol. 1, "Energy Levels of Nuclei, A = 5 to A = 257," Springer-Verlag, OHG, Berlin, 1961.)

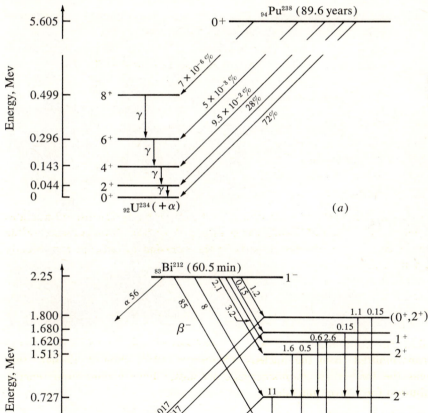

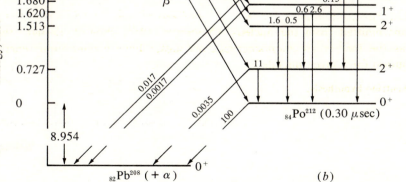

nucleus, because it is of order $(m_0/M'_D)T_e(\text{max})$, where T_e is the kinetic energy of the nuclear electron, we find that

$$T_e(\text{max}) = [M'_P - (M'_D + m_0)]c^2 \qquad (4\text{-}107)$$

In this equation, the primed masses are nuclear masses and m_0 is the rest mass of the electron.

FIGURE 4-18 Apparent nonconservation of linear momentum in the beta decay $He^6 \rightarrow Li^6 + e^-$. (By permission from J. Csikay and A. Szalay, *Proc. Intern. Congr. Nucl. Phys. Paris, 1958*, Publications Dunod, Paris, 1959.)

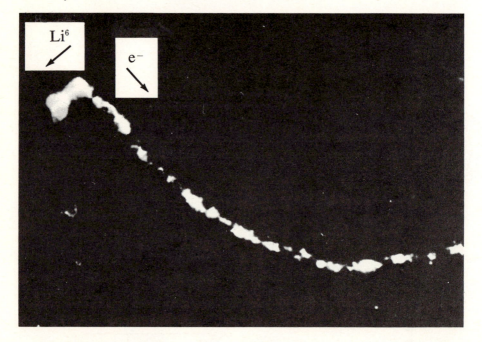

In addition to a possible violation of the energy conservation law by all electrons except those with maximum energy, there also appears to be a violation of the conservation law for angular momentum. We recall that on the basis of the neutron-proton hypothesis all odd-A nuclei are expected to have half-integral angular momenta (Sec. 1-2c). This is, indeed, the case. Since the emitted electron itself has spin $\frac{1}{2}$, it would change the angular momentum of the nucleus by this amount.[1] A decay of the type

$$_1H^3 \rightarrow {}_2He^3 + e^- \qquad (4\text{-}108)$$

[1] If the electron orbital angular momentum is also considered, the electron could carry off $(\frac{1}{2}, \frac{3}{2}, \frac{5}{2}, \ldots)$ $\hbar$ units of angular momentum.

would therefore require He^3 to have integral angular momentum, in contradiction to the facts.

Although investigated long after the neutrino hypothesis was put forward, beta decay seemed to violate also the law of conservation of linear momentum. Figure 4-18 shows a cloud-chamber photograph of the decay (from rest)

$$_2He^6 \rightarrow {}_3Li^6 + e^- \qquad (4\text{-}109)$$

in which the momentum vectors of the final products clearly do not add up to zero, as they should.

All difficulties concerning the conservation laws were overcome by the neutrino hypothesis of Pauli (1933). He proposed that another particle, besides the electron, is emitted in beta decay. To this particle, he assigned zero charge, zero or nearly zero mass (experimentally, the mass is known to be less than 1/2000 of the electron mass), and an intrinsic angular momentum $\frac{1}{2}\hbar$. The particle would carry off energy and linear momentum in accordance with Eqs. (2-9) and (2-10).

$$W = T = pc \qquad (4\text{-}110)$$

if the rest mass is exactly zero.

The beta decays (4-108) and (4-109) would then be of the form

$$_1H^3 \rightarrow {}_2He^3 + e^- + \bar{\nu} \qquad (4\text{-}111)$$
$$_2He^6 \rightarrow {}_3Li^6 + e^- + \bar{\nu} \qquad (4\text{-}112)$$

where, by definition, $\bar{\nu}$ is called an *antineutrino*.[1] A typical positron decay would be

$$_7N^{13} \rightarrow {}_6C^{13} + e^+ + \nu \qquad (4\text{-}113)$$

where ν is called a *neutrino*. With this assumption, the electron or positron kinetic energy T_e would be given by

$$T_e = [M'_P - (M'_D + m_0)]c^2 - W_{(\nu)} \qquad (4\text{-}114)$$

where $W_{(\nu)}$ is the energy (4-110) carried off by the antineutrino or neutrino.[2] Therefore, even though the mass difference for a given decay is fixed, the electrons can have a continuous energy distribution. Also, in the decays (4-111) to (4-113), angular and linear momentum can be balanced.

Although it may seem from the above discussion that the neutrino exists only to save the conservation laws of physics, its reality is now beyond doubt. The neutrino, having no charge (and no magnetic moment), does not produce

---

[1] Since the neutrino has a spin $\frac{1}{2}$, it is expected to obey the Dirac theory (see Sec. 3-4d). In addition to a neutrino, an antiparticle, called antineutrino, should also exist. We discuss the difference between these particles in Sec. 4-6g. Just as electron is a generic term used for positrons and negative electrons, neutrino is a term often used generically for a neutrino or an antineutrino. Usually no confusion results.

[2] This equation neglects the recoil kinetic energy of the daughter nucleus. See Eq. (4-115).

any ionization and therefore cannot be detected directly. Furthermore, it does not carry electric and magnetic fields like a photon and does not exert electromagnetic forces on an electron. But in an interaction with a nucleus, a neutrino can produce the inverse beta-decay reaction and this behavior has been detected (see Sec. 4-6g).

4-6b Energetics of beta decay.   Conservation of energy and momentum require

$$M_P' c^2 = M_D' c^2 + T_D + m_0 c^2 + T_e + W_{(v)} \qquad (4\text{-}115)$$

where nuclear masses are again primed, and

$$0 = \mathbf{p}_D + \mathbf{p}_e + \mathbf{p}_{(v)} \qquad (4\text{-}116)$$

where, in addition to the quantities mentioned previously, the recoil kinetic energy T_D and momentum $\mathbf{p}_D$ of the daughter nucleus are included. The order of magnitude of T_D can be shown to be $(m_0/M_D')(T_e + W_{(v)})$. For most purposes it is, therefore, negligible.

As in alpha decay, it is convenient to rewrite Eq. (4-115) in terms of atomic masses. Since the atomic number of the daughter nucleus is $Z_P + 1$ in electron decay and $Z_P - 1$ in positron decay, we find for electron decay

$$M_P c^2 = M_D c^2 + T_{e^-} + T_{\bar{v}} \qquad (4\text{-}117)$$

and for positron decay

$$M_P c^2 = M_D c^2 + 2m_0 c^2 + T_{e^+} + T_v \qquad (4\text{-}118)$$

In these expressions, it has been assumed that the neutrino has zero rest mass and that atomic electron binding-energy differences are negligible. The corresponding definitions of the Q values are

$$Q_{\beta^-} = T_{e^-} + T_{\bar{v}} = T_{e^-}(\max) \qquad (4\text{-}119)$$
$$Q_{\beta^-} = (M_P - M_D)c^2 \qquad (4\text{-}120)$$
$$Q_{\beta^+} = T_{e^+} + T_v = T_{e^+}(\max) \qquad (4\text{-}121)$$
$$Q_{\beta^+} = (M_P - M_D - 2m_0)c^2 \qquad (4\text{-}122)$$

where the first lines are definitions and the second lines follow from Eqs. (4-117) and (4-118). Negative electron decay is therefore possible if $M_{P(Z)} > M_{D(Z+1)}$; positron decay can occur only if $M_{P(Z)} > 2m_0 + M_{D(Z-1)}$. Another process, called electron capture, can always occur if $M_{P(Z)} > M_{D(Z-1)}$. We will discuss this in Sec. 4-6f.

Since beta-decay energies give directly mass differences of isobars, they can be used to construct mass parabolas (Fig. 2-15) and to test the predictions of the semiempirical mass formula. In the case of mirror nuclei, decay energies can yield values for the nuclear radius [Eq. (2-168) and Fig. 2-33]. Nuclear shell effects are also reflected in these energies.

4-6c Decay constant for beta decay.   Measured half-lives for beta decay vary approximately from 10^{-3} sec to 10^{16} years. As in gamma decay, we can classify various types of beta decay by the orbital angular momentum which the electron and neutrino carry off, and by the parity change which occurs. In addition, we can distinguish decays in which the intrinsic spins of the electron and neutrino are approximately parallel (*Gamow-Teller decays*) or antiparallel (*Fermi decays*). For the most common class of beta decays, the *allowed transitions* (in which zero *orbital* angular momentum is carried off), the decay constant increases roughly as the fifth power of the decay energy. These effects are explained in the theory of beta decay developed by Fermi (1934).

In an explanation of the beta-decay process, it is no longer possible to be guided by classical concepts, because one is now faced with the creation of two particles which did not preexist in the nucleus. The only classical theory concerned with a creation process is the emission of electromagnetic radiation from an accelerated charge. The rate of radiation [Eq. (4-47)] is determined by the specific nature of electric and magnetic fields and hence cannot be adapted directly to an *electron-neutrino field*. Nevertheless, Fermi developed a quantum theory of beta decay in analogy with the *quantum* theory of electromagnetic decay. The latter has to be considered briefly to gain the necessary background.

In the quantum mechanical treatment of a transition probability, we examine the entire system, which here would be the nucleus and the electromagnetic field that surrounds it. The transition takes the system from an initial state (excited nucleus + zero radiation) to a final state (final nucleus + radiation). We assume that only a very small disturbance is needed to effect the transition, so that no energy needs to be added to the system, and the transition occurs at a constant energy. The justification for this is the fact that radioactive decay times, i.e., lifetimes, are very long compared to nuclear periods. Effectively, the transition proceeds extremely slowly on a nuclear time scale, and the system is practically undisturbed on a nuclear time scale. In other words, the initial system needs to be only infinitesimally disturbed to make the transition proceed.

For convenience, the system is placed into a large closed box,[1] resulting in energy states which are enumerable (Sec. 2-2f). Inside this box the radiation field forms standing waves, each of which has a certain energy, shown schematically in Fig. 4-19a. In the initial state, the *excited nucleus + zero radiation* occupy only one definite energy level. The other levels are empty. One can show[2] from the complete Schrödinger equation (2-14) that if the system is subjected to a (time-dependent) potential of the form indicated in Fig. 4-19b, the system can make a transition to levels close by the initial level. Each of these levels corresponds to the *final nucleus + one photon*. The energy spread ΔE of the group levels which can be reached at time t after the perturbing potential ΔV is turned on, is approximately equal to $\hbar/t$ in accordance with the uncertainty principle.[3] As the time t increases ΔE tends to zero, so that energy conservation

---

[1] For simplicity, the box will be considered cubical, with a volume L^3. The nucleus is placed at the center of the box (see Sec. 2-2h).

[2] Schiff, 1955, sec. 29.

[3] If a system is subjected to any type of experimental condition or observation for a time t, the energy is uncertain by $\hbar/t$. Compare Sec. 4-3, where a system is observed for an effective time τ.

FIGURE 4-19 Quantum-mechanical treatment of transition probability. (*a*) Transition from initial state to a group of final states. Dashed levels are empty; solid levels are occupied. (*b*) Perturbing potential producing the transition.

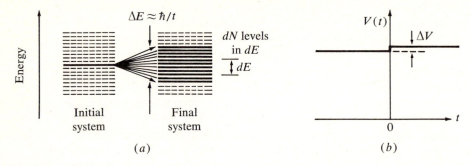

(*a*)   (*b*)

is finally obtained. Nevertheless, the decay constant (transition probability per unit time) is found to be proportional to dN/dE, the number of final states per unit energy. The complete expression for the decay constant is

$$\lambda = \frac{2\pi}{\hbar} \left| \int \psi_f^*(\text{system}) \, \Delta V \, \psi_i(\text{system}) \, dx \, dy \, dz \right|^2 \frac{dN}{dE} \qquad (4\text{-}123)$$

The integral extends over the volume of the box containing the system. Note that the wave functions refer to the entire system. In the case of gamma decay

$$\psi_i = \psi(\text{excited nucleus}) \qquad (4\text{-}124)$$

$$\psi_f = \psi(\text{final nucleus}) \, \psi(\text{photon}) \qquad (4\text{-}125)$$

As mentioned above, the photon wave function is assumed to form standing waves inside the closed box, so that conditions (2-81) and (2-83) apply to the wave vector **k** of the radiation. From this it is easy to calculate the *density of states* dN/dE. The wave function $\psi(\text{photon})$ is very similar to Eq. (2-112); to every set of integers n_x, n_y, n_z there belongs one state of the photon. To compute dN/dE, we use the fact that the length n of the radius vector in **n** space (Fig. 4-20) is directly proportional to the momentum p of the photon the photon

$$p = (p_x^2 + p_y^2 + p_z^2)^{\frac{1}{2}}$$

$$= \hbar(k_x^2 + k_y^2 + k_z^2)^{\frac{1}{2}}$$

$$= \frac{\hbar\pi}{L} (n_x^2 + n_y^2 + n_z^2)^{\frac{1}{2}}$$

$$= \frac{\hbar\pi}{L} n \qquad (4\text{-}126)$$

where Eqs. (2-27), (2-81), and (2-83) have been applied. The number of states dN

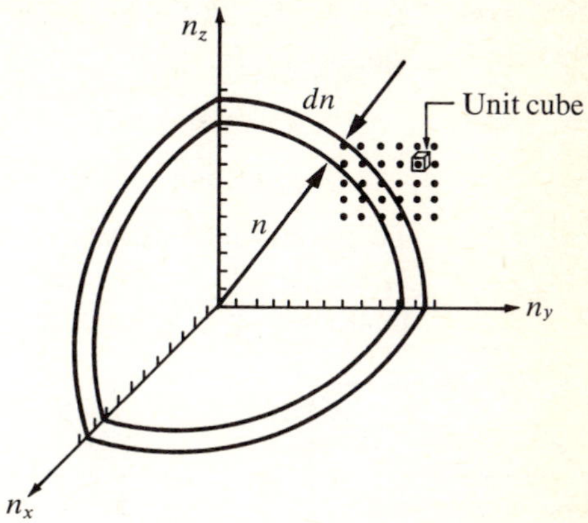

FIGURE 4-20 Volume in n-space. The length of
the radius vector is proportional to the momentum.
The number of states within a certain momentum
range is equal to the number of sets of positive
integers, n_x, n_y, n_z within the volume corresponding
to the momentum range.

corresponding to momenta lying between p and $p + dp$ is equal to the number
of sets of *positive* integers n_x, n_y, n_z lying between n and $n + dn$, where from
Eq. (4-126)

$$dn = \frac{L}{\pi\hbar} dp \qquad (4\text{-}127)$$

Since the volume in n space associated with each set of integers n_x, n_y, n_z is a
cube of unit volume, any volume in n space is numerically equal to the number
of sets of integers n_x, n_y, n_z within it. Therefore,

$$dN = \tfrac{1}{8}4\pi n^2 \, dn$$

$$= \frac{p^2 \, dp L^3}{2\pi^2 \hbar^3} \qquad (4\text{-}128)$$

and
$$\frac{dN}{dE} = \frac{p^2 (dp/dE) L^3}{2\pi^2 \hbar^3} \qquad (4\text{-}129)$$

For photons $E_r = p_r c$ [Eqs. (2-1) and (2-3)] so that

$$\frac{dN}{dE_r} = \frac{E_r^2 L^3}{\pi^2 c^3 \hbar^3} \qquad (4\text{-}130)$$

where a factor 2 has been included to take into account the two possible
directions of transverse polarization of electromagnetic radiation, which
represent independent states for the photons.

The calculation of the gamma decay constant requires evaluations of the
squared matrix element in Eq. (4-123). Note that the factor L^3 in Eq. (4-130) is
cancelled by a factor $1/L^3$ due to the normalization of ψ(photon) [compare Eq.
(2-112)]. Also, for electric dipole radiation, Eq. (4-55) is obtained after a complete evaluation of Eq. (4-123).

Proceeding now to beta decay, we can use the formalism of Eq. (4-123) by identifying for (negative) electron decay, e.g.,

$$\psi_i(\text{system}) = \psi(\text{parent nucleus}) \tag{4-131}$$

$$\psi_f(\text{system}) = \psi(\text{daughter nucleus})\,\psi(\text{electron})\,\psi(\text{antineutrino}) \tag{4-132}$$

To compute the density of states, i.e., to obtain enumerable states, the system is again placed at the center of a large box, as shown in Fig. 4-21. The number of final states per unit energy is the number dN_{tot} of electron-antineutrino states

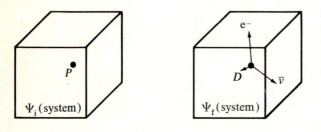

FIGURE 4-21   Beta decay (for negative electron emission). For computational purposes the system is placed into a large box of volume L^3. P is the parent and D the daughter nucleus.

in an energy range $dQ_{\beta-}$ [Eq. (4-119)]. Since for every electron state there is an independent set $dN_{\bar{\nu}}$ of available antineutrino states

$$dN_{\text{tot}} = dN_{e-}\,dN_{\bar{\nu}} \tag{4-133}$$

dN_{e-} and $dN_{\bar{\nu}}$ are each given by expression (4-128). From Eq. (4-123), the beta transition probability per unit time is equal to

$$\frac{2\pi}{\hbar}|\mathcal{M}|^2 \frac{dN_{\text{tot}}}{dQ_{\beta-}} \tag{4-134}$$

where, using Eqs. (4-131) and (4-132),

$$\mathcal{M} = \int \psi_D^*\psi_{e-}^*\psi_{\bar{\nu}}^*\,\Delta V\,\psi_P\,dx\,dy\,dz \tag{4-135}$$

The subscripts D and P refer to daughter and parent, respectively.

Expression (4-134) contains the square of a certain matrix element $\mathcal{M}$ and the density of states $dN_{\text{tot}}/dQ_{\beta-}$. Since the latter determines mainly the shape of a beta spectrum, it will be taken up first.

4-6d Shape of beta spectrum.   The evaluation of $dN_{\text{tot}}/dQ_{\beta-}$ depends on the type of experimental observation made. For example, if electrons are detected within a certain *fixed* momentum range δp_{e-} or energy range δT_{e-}, then from Eq. (4-119)

$$\frac{dN_{\text{tot}}}{dQ_{\beta-}} = \frac{dN_{\text{tot}}}{dT_{\bar{\nu}}}$$

$$= p_{e-}^2\delta p_{e-}p_{\bar{\nu}}^2\frac{dp_{\bar{\nu}}}{dT_{\bar{\nu}}}\frac{L^6}{4\pi^4\hbar^6} \tag{4-136}$$

using Eqs. (4-133) and (4-128). Relations (4-110) and (4-119)

give
$$\frac{dp_{\bar{\nu}}}{dT_{\bar{\nu}}} = \frac{1}{c} \tag{4-137}$$

and
$$p_{\bar{\nu}} = \frac{T_{e^-}(\max) - T_{e^-}}{c} \tag{4-138}$$

so that
$$\frac{dN_{\text{tot}}}{dQ_\beta} = p_e^2[T_e(\max) - T_e]^2 \delta p_e \frac{L^6}{4\pi^4 c^3 \hbar^6} \tag{4-139}$$

where the minus sign has been omitted from the subscripts because the same equation applies to positron decay. Substituting into expression (4-134), we find[1] for the probability per unit time $\Lambda(p_e)\delta p_e$ that an electron is emitted with momentum in the range p_e to $p_e + \delta p_e$

$$\Lambda(p_e)\delta p_e = p_e^2[T_e(\max) - T_e]^2 \delta p_e \frac{|\mathcal{M}|^2}{2\pi^3 c^3 \hbar^7} \tag{4-140}$$

We can also transform this into the probability per unit time that the kinetic energy of the electron lies in the range T_e to $T_e + \delta T_e$, by using Eqs. (2-9) and (2-10) to relate p_e and T_e. (In beta decay, electrons are usually emitted with relativistic speeds so that the nonrelativistic approximations are insufficient.)

$$\Lambda(T_e)\delta T_e = p_e(T_e + m_0 c^2)[T_e(\max) - T_e]^2 \delta T_e \frac{|\mathcal{M}|^2}{2\pi^3 c^5 \hbar^7} \tag{4-141}$$

The quantity $|\mathcal{M}|^2$ is practically independent of electron energy in the most common type of beta decay, but after evaluation is found to contain a coulomb penetration factor. Nonrelativistically, this is identical to Eq. (4-86) with γ given by the first term of Eq. (4-94).[2] For electrons $z = -1$, and for positrons $z = +1$. Therefore, the low-energy part of an electron distribution is enhanced; the electron is *held back* by the electric field of the nucleus. Positrons are *repelled* by the nucleus and the low-energy part of the energy spectrum is depleted. These effects are shown in Fig. 4-22. It is usual to separate the penetrability effects from $|\mathcal{M}|^2$ and to denote the penetrability (4-86) by $F(Z_D, p_e)$, called the *Fermi function*. Then

$$\Lambda(p_e)\delta p_e = F(Z_D, p_e)p_e^2[T_e(\max) - T_e]^2 \delta p_e \frac{|\mathcal{M}'|^2}{2\pi^3 c^3 \hbar^7} \tag{4-142}$$

where the prime on $\mathcal{M}'$ indicates that penetrability effects have been removed.

Equation (4-142) is the basis of the *Kurie plot* of beta spectra which is conveniently used to determine $T_e(\max)$. In a beta-ray spectrometer one obtains

---

[1] In all subsequent equations, the factor L^6 has been absorbed into $|\mathcal{M}|^2$ because it is cancelled by the normalization constants of ψ_{e^-} and $\psi_{\bar{\nu}}$.
[2] Since in the second term of Eq. (4-94) M_0 has to be replaced by the electron mass, the second term is negligible, as can be seen from the evaluations (4-97) and (4-98).

directly a quantity proportional to $\Lambda(p_e)$. Hence, the plot of

$$\frac{[\Lambda(p_e)/F(Z_D,p_e)]^{\frac{1}{2}}}{p_e} \tag{4-143}$$

against T_e yields a straight line if $\mathscr{M}'$ is independent of p_e. The straight line intersects the abscissa at $T_e = T_e\,(\text{max})$. Although beta-ray spectra are found in which $\mathscr{M}'$ is energy dependent, spurious deviations from a straight-line Kurie plot may occur. These might be caused by scattering and energy loss

FIGURE 4-22 Shapes of experimental beta-ray spectra. (a) Number of beta rays per unit momentum interval versus electron momentum. The units of momentum are gauss cm, because the product Br can be used to measure momentum [see Eq. (3.21)]. (b) Number of beta rays per unit energy interval versus kinetic energy. In both types of presentation the enhancement of the low-energy part of the β^- spectra and the depletion in the case of β^+ spectra are noticeable. [By permission from J. R. Reitz, *Phys. Rev.* **77**: 10 (1950), and Evans, 1955.]

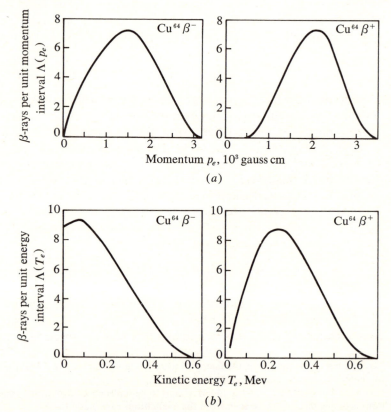

FIGURE 4-23 Kurie plot of the beta spectrum of
S^{35}. Curves A and B have been displaced vertically for
clarity. The curves show the effect of increasing source
thickness, which enhances the low-energy part of the
spectrum due to scattering and energy loss in the source
material. [By permission from R. D. Albert and C. S.
Wu, *Phys. Rev.* **74**: 847 (1948).]

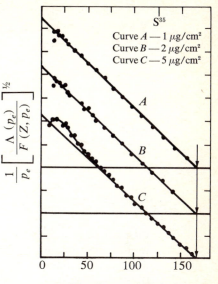

of electrons in the radioactive source (Fig. 4-23). The shape of the Kurie plot
near $T_e(\text{max})$ has also been used to set an upper limit to the neutrino mass.[1]

4-6e Lifetime and classification of beta decays.   The decay constant for beta
decay is obtained by integrating expression (4-142) over the entire spectrum

$$\lambda_\beta = \int_0^{p_e(\text{max})} \Lambda(p_e)\, dp_e$$

$$= \int_0^{\eta_0} F(Z_D,\eta)\eta^2(w_0 - w)^2\, d\eta\, |\mathscr{M}'|^2 \frac{m_0^5 c^4}{2\pi^3\hbar^7} \tag{4-144}$$

where for convenience a reduced momentum

$$\eta = \frac{p_e}{m_0 c} \qquad \eta_0 = \frac{p_e(\text{max})}{m_0 c} \tag{4-145}$$

and a reduced total energy

$$w = \frac{W}{m_0 c^2} = \frac{T_e}{m_0 c^2} + 1 \qquad w_0 = \frac{T_e(\text{max})}{m_0 c^2} + 1 \tag{4-146}$$

have been introduced. The integral (4-144) has been evaluated numerically. If
$\mathscr{M}'$ is energy independent, one finds

$$\lambda_\beta = f(Z_D,w_0)\, |\mathscr{M}'|^2 \frac{m_0^5 c^4}{2\pi^3\hbar^7} \tag{4-147}$$

The f function is plotted in Fig. 4-24. It is roughly proportional to $T_e^5(\text{max})$.

[1] Burcham, 1963, p. 605.

Although there exists no general estimate for $|\mathscr{M}'|^2$ analogous to the Weisskopf estimate for λ_γ [Eq. (4-64)], Eq. (4-147) is used in the classification of beta decays. It is usual to extract from $\mathscr{M}'$ the order of magnitude g of the interaction potential ΔV [Eq. (4-135)] and to rewrite Eq. (4-147) in the form

$$f(Z_D, w_0)t_{\frac{1}{2}} = \frac{t_0}{|M|^2} \qquad (4\text{-}148)$$

where

$$t_0 = \frac{2\pi^3\hbar^7 \ln 2}{m_0{}^5 c^4 g^2}$$

$$\approx 6{,}000 \text{ sec} \qquad (4\text{-}149)$$

$$M = \frac{\mathscr{M}'}{g} \qquad (4\text{-}150)$$

FIGURE 4-24 Plot of the f-function versus the maximum kinetic energy of the beta ray, for electron and positron emitters. Z is the atomic number of the daughter nucleus. For $T_e(\text{max}) > 1$ Mev, f is roughly proportional to the fifth power of $T_e(\text{max})$. [By permission from E. Feenberg and G. Trigg, *Rev. Mod. Phys.* **22**: 399 (1950), as adapted by Evans, 1955.]

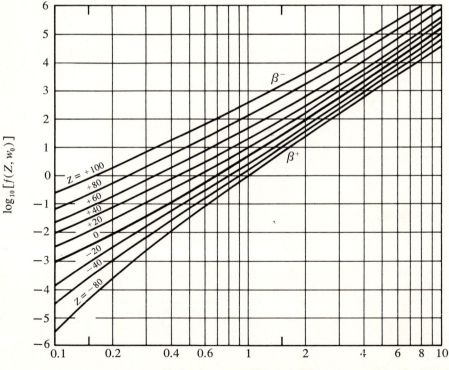

Equation (4-148) very neatly separates the kinematical factors in beta decay from the nuclear effects contained in M (or $\mathscr{M}'$). As in gamma decay, M is sensitive to selection rules and to the orbital momentum $\mathbf{L}_\beta \hbar$ carried off by the electron-neutrino pair. The magnitude of $|M|^2$ is decreased roughly by 10^{-2} to 10^{-4} for unit increase in L_β. Since the electron-neutrino pair also carries off intrinsic spin angular momentum $\mathbf{S}_\beta \hbar$, conservation of angular momentum requires

$$\mathbf{I}_P = \mathbf{I}_D + \mathbf{L}_\beta + \mathbf{S}_\beta \tag{4-151}$$

where, as previously, the subscripts P and D denote parent and daughter. One can show that the parity change in beta decay is $(-1)^{L_\beta}$, so that parity conservation requires

$$\pi_P = (-1)^{L_\beta} \pi_D \tag{4-152}$$

The quantity $|M|^2$ can be expanded in terms of increasing order in L_β. This is analogous to the expansion (4-63) for gamma decay.

$$|M|^2 = |M(L_\beta = 0)|^2 + |M(L_\beta = 1)|^2 + |M(L_\beta = 2)|^2 + \cdots \tag{4-153}$$

The *lowest* value of L_β satisfying Eqs. (4-151) and 4-152) will determine the dominant term in $|M|^2$ and thus the magnitude of $ft_{\frac{1}{2}}$ or λ_β. Decays with $L_\beta = 0$ are called *allowed*, with $L_\beta = 1$ *first forbidden*, $L_\beta = 2$ *second forbidden*, etc. We mentioned, in Sec. 4-6c, that decays with $S_\beta = 0$ (*Fermi decays*) are distinguished from those with $S_\beta = 1$ (*Gamow-Teller decays*), but the difference in $|M|^2$ for these two types of decay is not appreciable.

The selection rules (4-151) and (4-152) are applied to some specific examples in Table 4-3.

Experimental $\log ft_{\frac{1}{2}}$ values are shown in Fig. 4-25. Decays with the lowest values, clustering around 3.5, occur between mirror nuclei (Sec. 2-7). They are called *super-allowed* because the wave functions of the initial and final nuclei

TABLE 4-3 Examples of beta decays

Initial nucleus†		Final nucleus		Predominant decay mode
$_2\text{He}^6$	(0^+)	$_3\text{Li}^6$	(1^+)	Allowed, G.T.‡
$_8\text{O}^{14}$	(0^+)	$_7\text{N}^{14}*$	(0^+)	Allowed, F.‡
$_0\text{n}^1$	$(\frac{1}{2}^+)$	$_1\text{H}^1$	$(\frac{1}{2}^+)$	Allowed, G.T. and F. mixed
$_{16}\text{S}^{35}$	$(\frac{3}{2}^+)$	$_{17}\text{Cl}^{35}$	$(\frac{3}{2}^+)$	Allowed, G. T. and F. mixed
$_{39}\text{Y}^{91}$	$(\frac{1}{2}^-)$	$_{40}\text{Sr}^{91}$	$(\frac{5}{2}^+)$	First forbidden, G.T.‡
$_{17}\text{Cl}^{38}$	(2^-)	$_{18}\text{A}^{38}*$	(2^+)	First forbidden, G.T. and F. mixed
$_4\text{Be}^{10}$	(3^+)	$_5\text{B}^{10}$	(0^+)	Second forbidden, G.T.‡

† The total angular momentum and parity of each state is given.
‡ Only possible decay mode, called *unique transition*.

overlap so perfectly that the value of $|M|^2$ [Eq. (4-150)] is approximately equal to unity.[1] In a single-particle shell model we would assume that the last proton, for example, emits a positron and neutrino, and becomes a neutron. In a mirror nucleus, the neutron would have the identical wave function as the proton, so that there is full overlap between their wave functions. In most other nuclei, the

FIGURE 4-25 Frequency distribution of $\log ft_{\frac{1}{2}}$ values for the known beta emitters. (By permission from C. E. Gleit, C. W. Tang, and C. D. Coryell, "Beta-Decay Transition Probabilities," Nuclear Data Sheets, vol. 5, set 5, 1963.)

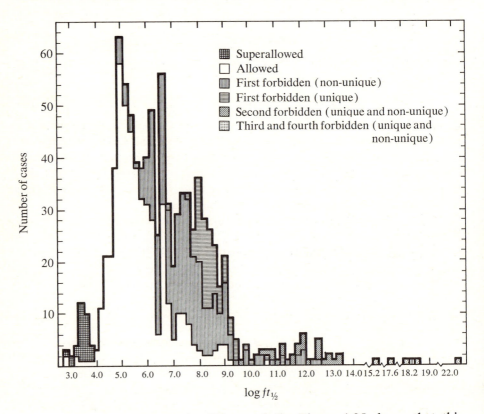

last proton and neutron are in different shells. Figure 4-25 shows that this decreases the decay probability by about 10^{-2}. The extreme sensitivity of beta-decay rates to the overlap of the initial and final nuclear wave functions causes the large spread among the various orders of beta decay apparent in Fig. 4-25. Also, this usually makes it impossible to recognize the order of a given beta decay solely from the $ft_{\frac{1}{2}}$ value. Independent information on spins and parities must be available and was, in fact, used to construct Fig. 4-25.

[1] For more accurate evaluations see Segrè, 1964, sec. 9-9.

4-6f Electron-capture decay.   Another process which is analogous to positron decay was discovered by Alvarez (1937). Under certain conditions, an atomic electron can be captured by a nucleus with the emission of a neutrino. The most probable capture is from the K shell because a K electron has the greatest probability of being inside the nucleus.[1]

FIGURE 4-26   Electron-capture process. In the example shown a K electron is captured, although capture can also take place from an outer atomic shell.

Initial state   Final state

The energetics of the process can be recognized from Fig. 4-26. In the initial state there is a parent atom. In the final state, there is an excited daughter atom plus a neutrino. Note that in the final state the nuclear charge and the number of atomic electrons still balance. From conservation of energy

$$M_P c^2 = M_D c^2 + E_B + T_\nu \qquad (4\text{-}154)$$

where E_B is the binding energy of the (missing) electron in the daughter nucleus. The atomic masses refer to the atoms in their ground states. The recoil energy of the daughter nucleus has been neglected. The Q value for electron capture is defined to be equal to the kinetic energy of the neutrino

$$Q_{\text{e.c.}} = T_\nu$$
$$= (M_P - M_D)c^2 - E_B \qquad (4\text{-}155)$$

Since for valence-electron capture $E_B \approx 0$, electron capture can always occur if $M_{P(Z)} > M_{D(Z-1)}$. Electron capture from an inner atomic shell is followed by a secondary process, the emission of x rays or Auger electrons (see Sec. 3-4c) by the daughter atom.

Figure 4-27 summarizes the energetics of the three beta-decay processes. All three types of beta decay can also lead to excited states. Typical decay schemes are presented in Fig. 4-28.

The decay constant $\lambda_{\text{e.c.}}$ for electron capture can be calculated by the same theory as presented previously. Because only one particle is emitted, expression (4-129) for the density of states applies. For allowed transitions $\lambda_{\text{e.c.}}$ turns out to be proportional to $T_\nu{}^2$ [compare Eq. (4-130)]. The ratio $\lambda_{\text{e.c.}}/\lambda_{\beta^+}$ is practically

[1] The electronic s wave function is finite at the origin. (Compare to the wave functions for a square well, Fig. 2-24.) In most cases, the probability of capture from the L shell is approximately 10 percent of that from the K shell. (Robinson and Fink, 1960.)

FIGURE 4-27 Energetics of beta-decay processes. (a) β^- decay. (b) β^+ and electron capture decay. β^+ decay is possible if $M_{P(Z)} - M_{D(Z-1)} > 2m_0$. (c) Only electron capture decay occurs if $0 < M_{P(Z)} - M_{D(Z-1)} < 2m_0$.

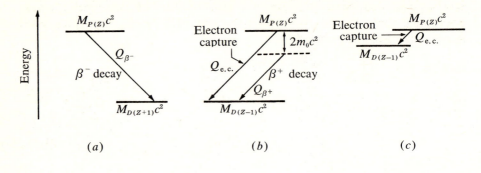

(a)　　　　　　　　　(b)　　　　　　　　　(c)

FIGURE 4-28 Typical decay schemes of beta emitters. Nuclear levels not populated in beta decay are shown in dotted lines. (These levels have been found by means of nuclear reactions.) Branching ratios are given in percent. Log $ft_{\frac{1}{2}}$ values are shown in brackets. (By permission from K. Way, A. Artna, and N. B. Gove, (eds.), "Reprint of Nuclear Data Sheets, 1959–1965," Academic Press Inc., New York, 1966.)

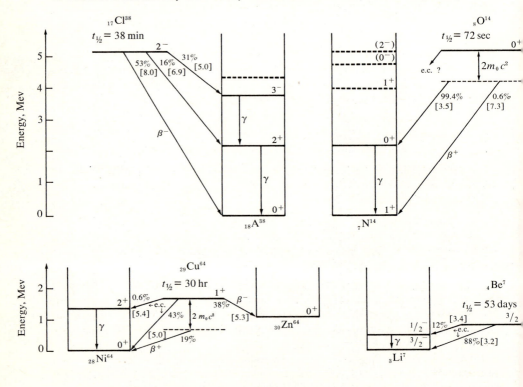

independent of any nuclear effects and so forms a good check of the theory of beta decay. Figure 4-29 shows the calculated ratio for the probability of K capture compared to positron emission. Good agreement with experiment has been found. Once the theory has been checked, the ratio $\lambda_{e.c.}/\lambda_{\beta+}$ can be used to estimate decay energies.

FIGURE 4-29 Probability of K-electron capture compared to positron decay, plotted versus end-point energy of the positron spectrum for various atomic numbers of the daughter nuclide. The plot applies to allowed spectra only. [By permission from E. Feenberg and G. Trigg, *Rev. Mod. Phys.* **22**: 399 (1950), adapted by Evans, 1955.]

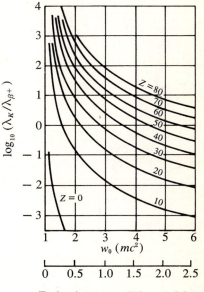

4-6g Inverse beta decay.   The theory of beta decay predicts that neutrinos should have a very small, but finite interaction probability with nuclei, about 10^{-19} times smaller than for ordinary nuclear reactions. We mentioned in Sec. 4-6a that such an interaction was found by Reines and Cowan (1953). They searched for the reaction

$$\bar{\nu} + p \rightarrow n + e^+ \qquad (4\text{-}156)$$

where the antineutrinos were produced by beta decays occurring in a nuclear reactor. This reaction is the *inverse* of neutron beta decay

$$n \rightarrow p + e^- + \bar{\nu} \qquad (4\text{-}157)$$

because in the sense of the Dirac theory (Sec. 3-4d) the creation of an electron is identical to the destruction of a positron. In other words, the process

$$n + e^+ \rightarrow p + \bar{\nu} \qquad (4\text{-}158)$$

is completely equivalent to beta decay.

In reaction (4-156), the creation of a neutron was signaled by detection of annihilation radiation from the positron, followed after several microseconds by the detection of slow-neutron capture gamma radiation (see Sec. 5-5c). Such a sequence of events selects reaction (4-156) from all possible backgrounds. By turning the antineutrino-producing[1] reactor on and off, the probability for producing the reaction could be checked and was found to be in agreement with theory.

In an analogous experiment, Davis (1955) tried to produce the inverse to the electron-capture process

$$_{18}A^{37} + e^- \rightarrow {}_{17}Cl^{37} + v \qquad (4\text{-}160)$$

but was unable to detect the reaction

$$Cl^{37} + v \rightarrow A^{37} + e^- \qquad (4\text{-}161)$$

adjacent to a reactor. Since a reactor produces antineutrinos,[1] this clearly demonstrates that neutrinos and antineutrinos are different particles. We now know that they are distinguished by the direction of their intrinsic spin: Neutrinos have their intrinsic spins antiparallel to the direction of travel; antineutrinos have their intrinsic spins parallel to the direction of travel [Goldhaber, Grodzins, and Sunyar (1958)].

4-6h Parity nonconservation in beta decay.   The property of neutrinos just described does not conserve parity and leads to a certain type of nonconservation of parity in beta decay, in agreement with experiment. Consider a neutrino traveling to the right as shown in Fig. 4-30a. Its intrinsic spin angular momentum $s\hbar$ points to the left. If this property of the neutrino would conserve parity, the mirror situation, Fig. 4-30b, would also be possible: the neutrino would travel to the left, but its spin $s\hbar$ would also point to the left because the angular

FIGURE 4-30 Parity nonconserving property of neutrino. (a) Neutrino traveling to right with a velocity v_v. (b) Experiment as seen in mirror; this situation does not occur in nature for a neutrino.

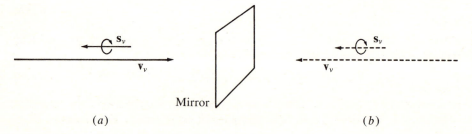

(a)   (b)

---

[1] Neutron capture on stable nuclides usually produces beta-radioactive nuclides, which emit antineutrinos.

momentum vector does not change direction in this mirror experiment.[1] This situation, however, cannot occur for the neutrino, because, as stated above, all neutrinos have their spins antiparallel to their velocity vectors. The situation depicted in Fig. 4-30b corresponds to an *anti*neutrino. In other words, the mirror experiment is possible only if the particle is changed to an antiparticle.

This parity nonconserving property of neutrinos was first found in a series of experiments suggested by Lee and Yang (1956). They predicted on the basis of certain meson decays that mirror experiments in beta decay do not occur in nature, unless every one of the light particles (electron, antineutrino) is changed into its antiparticle (positron, neutrino). In particular, two parity nonconserving effects were predicted and found: (1) The angular distribution of beta rays from polarized nuclei is not symmetrical about a plane through the nucleus, perpendicular to the axis of polarization. (2) Electrons emitted in beta decay have their spins preferentially antiparallel to their direction of travel, and the opposite occurs for positrons. That these two properties do not conserve parity can be seen by considerations similar to those described in connection with Fig. 4-30.[2]

Parity nonconserving effects occur only for the light particles in beta decay. Nuclear states have a definite parity with extremely high precision so that selection rule (4-152) must be obeyed. Also, ordinary electrons do not have any preferential direction of polarization.

4-6i Nuclear structure information from beta decay. Part of the nuclear structure information from beta decay is similar to that obtained from gamma decay. The decay energies are useful to check energy systematics in nuclei, particularly the semiempirical mass formula and shell effects. The matrix element $\mathscr{M}$ [Eq. (4-135)] is very sensitive to the overlap between the parent and daughter wave functions. In mirror nuclei the overlap is practically perfect; this gives added support to the charge *symmetry* of nuclear interactions (Sec. 2-7). Also in mirror triads (see e.g., Fig. 2-34), beta transitions between corresponding levels, where permitted energetically, are superallowed. This indicates perfect overlap between the wave functions, and supports charge *independence* of nuclear forces (Sec. 2-7).

The interaction potential ΔV [Eq. (4-135)] also contains valuable information, which will provide new insight into the role played by various mesons within the nucleus. At present, only some general limitations on ΔV required by symmetry and by parity nonconserving effects are understood. The magnitude g of the interaction potential has not yet been explained. Empirically it is approximately 10^{-6} times as small as nuclear interactions or 10^{-4} times as small as electromagnetic interactions. The beta-decay interaction is therefore called a *weak interaction*. Presumably, it is caused by a force field which is not nuclear, electromagnetic, or gravitational.

---

[1] If we assume the neutrino angular momentum can be treated as that of a rotating body, we can see that the sense of rotation is not altered by the mirror experiment proposed in Fig. 4-30.
[2] Burcham, 1963, sec. 16-7.

PROBLEMS

4-1   A radioactive source consists of a mixture of two radioactive nuclides whose initial activities are identical. One nuclide decays with a half-life of $\frac{1}{2}$ year, the other with a half-life of $\frac{1}{3}$ year. What fraction of the initial activity remains after 1 year?

4-2   The natural abundance of U^{235} is 0.72 percent and of U^{238} 99.3 percent. Assuming that in the process of element formation, both isotopes had been formed with equal abundance, what can we infer about the time at which the elements were formed? The half-lives of U^{235} and U^{238} are 6.8×10^8 and 4.6×10^9 years, respectively.

4-3   A radioactive sample produced in a nuclear reactor has the following decay curve:

Time of observation, min	Activity, counts/sec	Time of observation, min	Activity, counts/sec
0	366	10	128
1	289	15	99
2	241	20	78
3	210	25	63
4	189	30	50
5	173	35	42
6	161	40	35
7	151	45	30

After several hours a constant background of 15 counts/sec was observed. Compute the half-lives and relative initial activities of the radioactive isotopes in the sample.

4-4   The radioactive nuclide Na^{24} ($t_{\frac{1}{2}} = 14.8$ hours) can be produced by neutron bombardment of Na^{23}. If the production rate of Na^{24} is 10^8/sec, and the bombardment is started with a fresh sample of Na^{23}, compute   **(a)** The maximum activity of Na^{24} (in curies) which could be produced.   **(b)** The bombardment time needed to produce 90 percent of the maximum activity.   **(c)** The number of radioactive atoms of Na^{24} left 3 hours after the bombardment (b) was stopped.

4-5   **(a)** One gram of the element potassium emits 29 β^- particles/sec due to the decay of K^{40} (natural abundance 0.012 atom percent). Because of competing electron-capture decay to an excited state of A^{40}, gamma-rays are also emitted. One finds $N_\gamma/N_\beta = 0.12$. There is no electron-capture decay to the ground state of A^{40}. What is the half-life of K^{40}?   **(b)** This radioactive decay can be used to determine the geological age of a mineral by measuring the concentration ratio of K^{40} to A^{40}. We have to assume that all A^{40} present is from the decay of K^{40} and that no A^{40} has escaped from the mineral. What is the age of a mineral for which $N(A^{40})/N(K^{40}) = 0.5$?

4-6   (a) A radioactive parent nuclide has a radioactive daughter. Find an expression for the time at which the activity of the daughter is a maximum, assuming an initially pure radioactive parent. What is the ratio of the activity of the parent to that of the daughter at that time? (b) An initially pure sample of Th^{227} ($t_{\frac{1}{2}} = 18.2$ days) decays into Ra^{223} ($t_{\frac{1}{2}} = 11.7$ days) by alpha emission. Ra^{223} is also an alpha emitter; when is its activity at a maximum? (c) What will be the ratio of the activity of Th^{227} to the activity of Ra^{223} after several months?

4-7   Radium (Ra^{226}, $t_{\frac{1}{2}} = 1622$ years) decays into radon (Rn^{222}, $t_{\frac{1}{2}} = 3.82$ days), which is a monoatomic gas. A sample of 100 mC initially pure radium is allowed to come to equilibrium with its decay products for several months. Compute the number of radon atoms present at that time. What volume would this gas occupy at NTP?

4-8   An initially pure sample of 10 mC Ra E(Bi^{210}, $t_{\frac{1}{2}} = 5.01$ days) is allowed to decay to polonium (Po^{210}, $t_{\frac{1}{2}} = 138.4$ days). Compute the maximum activity of the polonium.

4-9   The nuclide Co^{60} has an isomeric level as shown. (a) What are the separate decay constants for β, e^-, and γ decay? (b) What is the appropriate Weisskopf estimate for the gamma decay constant? (c) What is the width of the isomeric state of Co^{60} (in ev)?

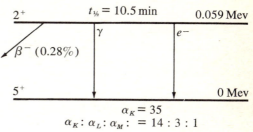

4-10   In B^{11}, a gamma transition leads from a level at 9.28 Mev (spin-parity $\frac{5}{2}^+$) to a level at 4.46 Mev (spin-parity $\frac{5}{2}^-$). What is the dominant multipolarity of the transition? Calculate the expected gamma width (in ev) on the basis of the Weisskopf model. (The measured value of Γ_γ is equal to 5.4 ev.)

4-11   In Zr^{90}, an isomeric level at 2.315 Mev has a half-life of 0.83 sec. The level has a 16 percent branch to a state at 2.182 Mev and a 84 percent branch to the ground state. The two transitions have K conversion coefficients of 2.0 and of 4×10^{-4}, respectively. Use Eqs. (4-69) and Fig. 4-9 to suggest multipolarities for the transitions and possible spin and parity assignments for the levels.

4-12   The beta decay of $_{55}Cs^{137}$ leads to an isomeric state of $_{56}Ba^{137}$, which decays by a transition of 0.6616 Mev. Compute the energies of the K and L conversion electrons. (The K and L binding energies for Cs are 35.9 and 5.7 kev and for Ba 37.4 and 6.0 kev, respectively.)

4-13   In Fig. 4-17a, the alpha-decay energy of Pu^{238} is given, as well as energies of excited states of U^{234}. Compute the kinetic energy of the alpha-particle group that leads to the 0.499-Mev excited state of U^{234}.

4-14 (a) Show from the semiempirical mass formula that for a given isotope the slope of the alpha-decay energy versus neutron number should be negative. (b) Compute values of the slope for $Z = 86$ and $N = 120, 130$, using any consistent set of energy parameters. Compare with Fig. 4-12.

4-15 (a) Show from the semiempirical mass formula that for a given isotone the slope of the alpha-decay energy versus atomic number should be positive. (b) Compute the slope for $N = 120$, near $Z = 84$, using any consistent set of energy parameters. Compare with Fig. 4-12.

4-16 The nuclide Nd^{144} emits 1.83-Mev alpha particles. Use (Eq. 4-95) to estimate its half-life. (The experimental half-life is 2.4×10^{15} years.)

4-17 Assume that two isotopes of the element $_{98}Cf$ ($A \approx 250$), differing in neutron number by 4, have identical alpha-decay energies. (a) Which isotope should have the longer half-life? (b) Compute the expected percentage difference in the half-lives.

4-18 Calculate the end-point energy of the beta spectrum of the neutron (in Mev) from the neutron and the hydrogen masses.

4-19 Show that if the end-point energy of a beta emitter is much less than m_0c^2, the ratio of the average beta energy to the endpoint energy is $\frac{1}{3}$. Assume that the Fermi function is approximately constant over the energy interval in question.

4-20 Assume that in Fig. 4-18 the angle between the recoiling Li^6 nucleus and the electron is $90°$. From this fact alone show that the kinetic energy of the electron shown must be less than one-half of the maximum available energy. (You may find a graphical solution useful.)

4-21 Assume that in the decay shown in Fig. 4-18 the angle between the recoiling Li^6 nucleus and the electron is $90°$ and that the Li^6 nucleus has the same momentum magnitude as the electron. Compute the kinetic energies of the three product particles of this decay (Li^6, e^-, $\bar{\nu}$). The decay energy of He^6 is 3.57 Mev.

4-22 Show by use of the uncertainty principle that the kinetic energy spread of electrons inside the C^{12} nucleus is large compared to the largest beta-ray end-point energies found in light nuclei (≈ 15 Mev). (Use relativistic expressions, but you may assume that for the hypothetical electron inside the nucleus $p \gg m_0c$.) This discrepancy is sometimes used as an argument against the existence of electrons in nuclei.

4-23 Assuming a Be^7 nucleus is at rest before a K-capture takes place, what velocity (in cm/sec) and what energy (in ev) are imparted to it after the K-capture process, assuming the neutrino has zero rest mass? (The atomic mass difference between Be^7 and Li^7 is 0.86 Mev.)

4-24 A 2.57-cm^3 ampule containing tritium gas at NTP was found to produce heat at a rate of 0.1909 cal/hour. The half-life of tritium is 12.46 years and its beta spectrum has an end-point energy of 19.4 kev. Compute: (a) The activity of the tritium sample in curies. (b) The average energy of the beta particles emitted. (c) The ratio of the average energy to the maximum energy (see Prob. 4-19).

4-25 In the following table, certain initial and final nuclear states are given. State whether natural radioactive decay between these states is permitted, and if so, describe the dominant decay mode as accurately as you can. Assume that always $(M_{\text{initial}} - M_{\text{final}})c^2 = 2$ Mev.

	Initial nucleus				Final nucleus			
	A	Z	I	parity	A	Z	I	parity
(1)	A	Z	1	−	A	Z	1	+
(2)	A	Z	$\frac{3}{2}$	+	A	Z	$\frac{7}{2}$	+
(3)	A	Z	0	−	A	Z	0	−
(4)	A	Z	5	+	A	Z	4	+
(5)	A	Z	3	−	A	Z − 1	3	−
(6)	A	Z	$\frac{9}{2}$	+	A	Z − 1	$\frac{1}{2}$	−
(7)	A	Z	0	−	A	Z + 1	0	−
(8)	A	Z	2	+	A	Z + 1	1	−
(9)	A	Z	0	+	A	Z − 2	2	+
(10)	A	Z	0	+	A − 4	Z − 2	2	+

4-26 The figure below shows the decay scheme of Br^{80}. **(a)** Use the information given and Fig. 4-24 to compute the log-*ft* values of the three beta decays. **(b)** Use Fig. 4-29 to compute the expected K/β^+ ratio for the positron branch. **(c)** Classify the various branches by order and type.

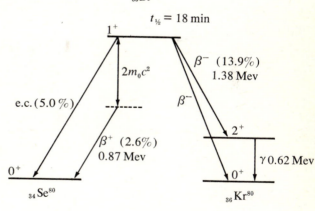

NUCLEAR REACTIONS ░░░░░░░░░░░░░░░░░░░░░░░░5

5-1 INTRODUCTION

Between 1919, when Rutherford announced the discovery of the artificial nuclear transmutation[1]

$$\text{He}^4 + \text{N}^{14} \rightarrow \text{H}^1 + \text{O}^{17} \tag{5-1}$$

and 1939, when fission was discovered (Hahn and Strassman, Meitner and Frisch), nearly all known nuclear processes which could be initiated with bombarding energies up to approximately 10 Mev were found. Since then, bombarding energies have been extended to roughly 10 Bev, and many new types of reactions have been produced, particularly those involving mesons and other unstable particles. Although it is now clear that mesons play a fundamental role in nuclear forces, the present discussion is limited to nuclear reactions below the threshold for meson production (≈ 150 Mev).

---

[1] A cloud-chamber photograph of this reaction is shown in Fig. 5-7.

171

Detailed theories of nuclear reaction were patterned after the two, apparently contradictory, models of nuclear structure mentioned in Chap. 2: the liquid-drop model and the shell model. In one theory, it was assumed (Bohr, 1936) that a nuclear projectile incident on a nucleus would interact strongly with all the nucleons in the nucleus and quickly share its energy with them. The compound nucleus so created would decay in a manner independent of its mode of formation. In the reaction theory based on the shell model (Bethe, 1940; Fernbach, Serber, and Taylor, 1949; Feshbach, Porter, and Weisskopf, 1954), it was proposed that an incident nucleon would interact with the nucleus via the shell-model potential and that the probability of absorption into the compound nucleus would be relatively small. These different aspects of a nuclear reaction can be unified into a single theory (Weisskopf, 1957; Feshbach, 1958).

FIGURE 5-1 Sequence of stages in a nuclear reaction according to Weisskopf. (By permission from Weisskopf, 1957.)

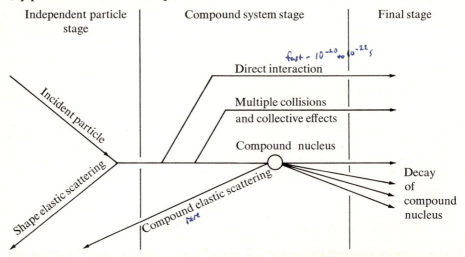

According to Weisskopf, any nuclear reaction proceeds through a series of stages, indicated schematically in Fig. 5-1. When the incident particle reaches the edge of the nuclear potential, the first interaction will be a partial reflection of the wave function, called *shape elastic scattering*. We recall that any potential discontinuity has a finite reflection coefficient for an incident wave [see Eq. (2-162)] which is independent of the direction of travel of the wave. The part of the wave function which enters the nucleus undergoes absorption. Feshbach proposes that the first step in the absorption process consists of a two-body collision. In other words, if the incident particle is a single nucleon, it interacts with a single nucleon in the nucleus and raises it to an unfilled level as shown in Fig. 5-2. If the struck nucleon leaves, a *direct reaction* occurs. Presumably this process becomes more probable at higher energies because, then, at least one nucleon would have a good chance of receiving enough energy to leave the nucleus.

If the struck nucleon does not leave the nucleus, more complicated interactions can set in. The incident nucleon (or the struck nucleon) may interact with a second nucleon in the nucleus, in turn raising it to an unfilled level. Under proper conditions the nucleus could be excited to a collective state (Sec. 2-5d), and one of the nucleons could leave. If this does not occur, each of the (three) nucleons which are now in unfilled levels in the nucleus can interact with other nucleons until finally the energy sharing envisaged by the compound-nucleus theory has occurred.

FIGURE 5-2 First step in a nuclear reaction according to the unified reaction theory of Feshbach. The incident particle collides with one nucleon in the nucleus and raises it to a higher state. If the nucleon leaves, a *direct reaction* occurs (shown in dotted lines). Neutrons and protons are not distinguished in the diagram.

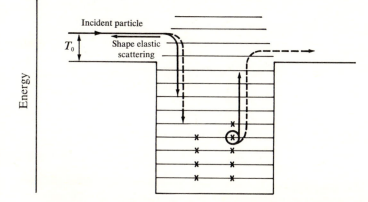

The compound nucleus is formed in such a complicated set of interactions that it probably does not "remember" details of the initial stage of formation. Hence, its decay should be independent of the way it was produced. It may happen that the incident particle (or a particle of the same kind as the incident particle) is emitted by the compound nucleus with the same (c.m.) energy as the incident particle. This is called *compound elastic scattering*. The particle so emerging cannot be distinguished from the shape elastically scattered particle, except possibly by a slight time delay.

We can also understand from this picture of a nuclear reaction that as the incident (c.m.) energy T_0 is varied (Fig. 5-2) it may exactly correspond to a virtual level of the nuclear potential (Fig. 2-29). The probability of finding the incident particle in the nucleus is then high and the nuclear reaction probability has a *potential* or *single-particle* resonance. It is more difficult to see that the probability of compound-nucleus formation also has many resonances (*compound-nucleus resonances*). We can only note that any quantum-mechanical system of high excitation[1] has many close-lying levels because, then, many different modes of excitation can occur with similar excitation energies. At present, both the energy and the width of compound-nucleus resonances can be expressed only in terms of empirical constants (see Sec. 5-5), but theoretical understanding is developing.

Certain aspects of nuclear reactions are independent of the detailed interaction mechanism and can be derived from conservation of energy, linear momentum, and angular momentum. Parity is also conserved to an extremely high degree. Further, the number and kind of nucleons in any reaction is constant until reaction energies are high enough to create nucleon-antinucleon pairs.[2] We will discuss the application of these conservation laws to nuclear reactions before considering details of the probability of interaction or cross section.

5-2 APPLICATION OF CONSERVATION LAWS

For bombarding energies below 100 Mev, nuclear reactions usually produce two products, i.e., they are of the type

$$a + X \rightarrow b + Y \tag{5-2}$$

where
a = bombarding particle $(beam)$
X = target (at rest in the lab. system)
b = light reaction product
Y = heavy reaction product

To shorten the notation a reaction of the type (5-2) is designated by

$$X(a,b)Y \tag{5-3}$$

Commonly, one reaction product is light and the other heavy because of the binding energies of the nuclei involved. In some cases b and Y have comparable masses (*spallation reaction* or *fission*), or are identical. If b is a gamma ray, we speak of a *capture reaction* in which Y is the compound nucleus.

In most cases in which more than two products appear, it is possible to describe the process as a rapid sequence of two-product reactions

$$a + X \rightarrow b_1 + Y_1$$
$$Y_1 \rightarrow b_2 + Y_2$$
$$Y_2 \rightarrow b_3 + Y_3, \tag{5-4}$$
$$\cdots$$

[1] The excitation energy of the compound nucleus is $S + T_0$, where S is the separation energy of the incident particle from the ground state of the compound nucleus (see Fig. 5-22). S is of the order of 8 Mev for protons and neutrons.
[2] If antinucleons are counted negatively, the total number of nucleons in all known reactions is conserved.

Reaction (5-1) is an example of the type (5-2). Note that the number of neutrons and protons is conserved. Presently the number of known reactions is in the thousands.

5-2a Energetics. Conservation of linear momentum. Since the number of protons remains unchanged in a reaction, all masses can be written as *atomic* masses if electron binding-energy differences of a few ev are ignored. Conservation of energy, therefore, gives for the reaction (5-2)

$$M_a c^2 + T_a + M_X c^2 = M_b c^2 + T_b + M_Y c^2 + T_Y \qquad (5\text{-}5)$$

where T represents the (lab.) kinetic energy of each particle. The masses of a and X are ground-state masses. On the other hand, many reactions leave Y in excited states; in that case, M_Y represents the total mass energy of that state.

The Q value of the reaction is defined as the difference between the final and initial kinetic energies [compare Eq. (4-77)]

$$Q = T_b + T_Y - T_a \qquad (5\text{-}6)$$
$$Q = [M_a + M_X - (M_b + M_Y)]c^2 \qquad (5\text{-}7)$$

If Q is positive, the reaction is said to be *exoergic*; if Q is negative, it is *endoergic*. A reaction cannot take place unless particles b and Y emerge with positive kinetic energies, that is, $T_b + T_Y \geq 0$ or

$$Q + T_a \geq 0 \qquad (5\text{-}8)$$

Although this condition is necessary, it is not sufficient.

The Q value is an important quantity in a nuclear reaction. It can be determined from mass spectroscopy [Eq. (5-7)] or by measuring kinetic energies [Eq. (5-6)]. We can show, as a result of linear momentum conservation, only T_b and the angle θ of b with respect to the direction of a (Fig. 5-3) need to be determined. In the lab. system

$$M_a v_a = M_Y v_Y \cos \phi + M_b v_b \cos \theta$$
$$0 = M_Y v_Y \sin \phi - M_b v_b \sin \theta \qquad (5\text{-}9)$$

FIGURE 5-3 Nuclear reaction in the lab. system. (*a*) Initial situation. (*b*) Final situation.

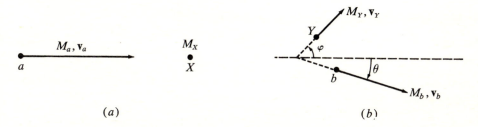

(*a*)   (*b*)

In order to eliminate ϕ, substitute $Mv = (2MT)^{\frac{1}{2}}$ for each particle and rewrite the equations

$$(M_a T_a)^{\frac{1}{2}} - (M_b T_b)^{\frac{1}{2}} \cos \theta = (M_Y T_Y)^{\frac{1}{2}} \cos \phi$$

$$(M_b T_b)^{\frac{1}{2}} \sin \theta = (M_Y T_Y)^{\frac{1}{2}} \sin \phi \qquad (5\text{-}10)$$

Squaring both equations and adding

$$M_a T_a - 2(M_a T_a M_b T_b)^{\frac{1}{2}} \cos \theta + M_b T_b = M_Y T_Y \qquad (5\text{-}11)$$

Eliminating T_Y with the help of Eq. (5-6)

$$Q = T_b\left(1 + \frac{M_b}{M_Y}\right) - T_a\left(1 - \frac{M_a}{M_Y}\right) - \frac{2}{M_Y}(M_a T_a M_b T_b)^{\frac{1}{2}} \cos \theta \qquad (5\text{-}12)$$

This is called the *Q equation*. Special cases of interest are those with $\theta = 90°$ and those with zero bombarding energy T_a. The latter reaction is possible only with neutrons, since the coulomb barrier prevents nuclear reactions with zero-energy charged particles. Energetically, the situation is then similar to Eq. (4-80).

Part of the incident energy T_a is used up as *kinetic energy of the center of mass* and is not available for the nuclear reaction itself. Although we can study all the resulting effects by means[1] of Eq. (5-12), more insight is gained if we consider the reaction in the c.m. system,[2] shown in Fig. 5-4 (see also Fig. 3-11). The kinetic energy *of* the center of mass is

$$T_{\text{c.m.}} = \tfrac{1}{2}(M_a + M_X)v_0^2 \qquad (5\text{-}13)$$

where $v_0 = v_a M_a/(M_a + M_X)$ is the speed of the center of mass. The kinetic energy T_0 of the initial particles *in* the c.m. system can be calculated in two equivalent ways as

$$T_0 = T_a - T_{\text{c.m.}} \qquad (5\text{-}14)$$

or $$T_0 = \tfrac{1}{2}M_a V_a^2 + \tfrac{1}{2}M_X V_X^2 \qquad (5\text{-}15)$$

where V represents the speed of each particle in the c.m. system (Fig. 5-4). Equations (5-14) and (5-15) both yield

$$T_0 = \frac{M_X}{M_a + M_X} T_a \qquad (5\text{-}16)$$

The *energy available for the nuclear reaction* is

$$Q + T_0 \qquad (5\text{-}17)$$

[1] R. D. Evans, 1955, chap. 12, sec. 2.
[2] Since the mass of the system changes from $M_a + M_X$ to $M_b + M_Y$, the c.m. system is not identical for the initial and final products. As long as all velocities are nonrelativistic and the fractional mass difference is small, this effect can be ignored. If it cannot be ignored, it is more useful to define a center-of-momentum system, which does not change during the reaction.

FIGURE 5-4 Nuclear reaction in c.m. system. (Compare Fig. 3-11.) (*a*) Initial situation. (*b*) Final situation. The speed of the center of mass is $v_0 = v_a M_a / (M_a + M_X)$. Also $M_b V_b = M_Y V_Y$.

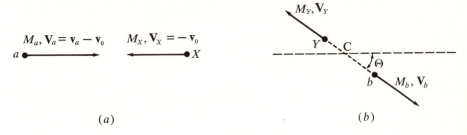

$$M_a, \mathbf{V}_a = \mathbf{v}_a - \mathbf{v}_0 \qquad M_X, \mathbf{V}_X = -\mathbf{v}_0$$

(*a*) (*b*)

which is equal to the kinetic energy of the reaction products in the c.m. system

$$Q + T_0 = \tfrac{1}{2} M_b V_b{}^2 + \tfrac{1}{2} M_Y V_Y{}^2 \tag{5-18}$$

This is easy to see, because if $T_{\text{c.m.}}$ is added to both sides of Eq. (5-18), the result is identical to Eq. (5-6).

A necessary and sufficient condition that the reaction proceed is that the right-hand side of Eq. (5-18) be positive, i.e.,

$$Q + T_0 \geq 0 \tag{5-19}$$

This would automatically satisfy Eq. (5-8). Using Eq. (5-16), the same condition is

$$T_a \geq \frac{-Q(M_a + M_X)}{M_X} \tag{5-20}$$

In the case of an endoergic reaction ($Q < 0$), Eq. (5-20) gives the threshold energy of the reaction. The threshold energy can also be derived by noting that at threshold, particles b and Y both move with the speed v_0 in the lab. system

$$(T_b + T_Y)_{\text{thresh}} = \tfrac{1}{2}(M_b + M_Y)v_0{}^2 \tag{5-21}$$

After a short calculation, using $M_b + M_Y \approx M_a + M_X$, Eq. (5-20) is obtained.

We can return to the lab. system from Fig. 5-4*b* by adding the velocity $\mathbf{v}_0$ to the velocities shown. Numerous interesting situations can then be examined geometrically.[1] For example, in the case of an endoergic reaction, particles b can appear with two different kinetic energies at the same lab. angle θ, if T_a is only slightly above the threshold energy. This occurs because T_a determines only the c.m. speed V_b [see Eq. (5-18) and note that $M_b V_b = M_Y V_Y$] and not the direction of the velocity $\mathbf{V}_b$. As shown in Fig. 5-5, under suitable conditions

[1] See Prob. 5-1.

FIGURE 5-5   Velocity diagram for particle b at an energy slightly above threshold of an endoergic reaction. At certain lab. angles θ, particle b appears with two different kinetic energies $\frac{1}{2}M_b v_b^2$ and $\frac{1}{2}M_b v_b'^2$.

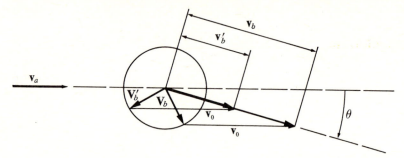

two different velocities $\mathbf{V}_b$ and $\mathbf{V}_b'$ of same magnitude can produce particles b at a given lab. angle θ with different speeds v_b and v_b'.

It is convenient to plot the c.m. mass-energy information for a nuclear reaction on a diagram similar to Fig. 5-6. The example illustrates an endoergic reaction ($Q < 0$).

A famous cloud-chamber photograph[1] of the (endoergic) nuclear reaction (5-1) $N^{14}(\alpha,p)O^{17}$ is shown in Fig. 5-7. The alpha particles were emitted by a *thorium active deposit*[2] which produces principal alpha groups of 8.8 Mev from Po^{212} (Th C$'$) and 6.1 Mev from Bi^{212} (Th C). From the path length of the inter-acting alpha particle (Fig. 5-7), its energy at the place of collision was calculated

---

[1] Blackett and Lees, 1932.
[2] A partial decay scheme is reproduced in Fig. 4-17b. Note that values of Q_α, and not T_α, are given. For more details see Evans, 1955, p. 516, fig. 1-3.

FIGURE 5-6   Center-of-mass energetics of a nuclear reaction. The illustration applies to an endoergic reaction, that is, $Q = (M_a + M_X)c^2 - (M_b + M_Y)c^2 < 0$. The compound-system mass energy is also indicated.

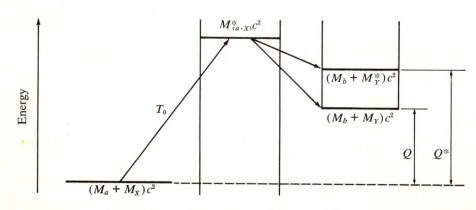

to be 3.9 Mev. The angle and range of the emitted proton allowed Q to be computed from Eq. (5-12), with the result $Q = -1.2$ Mev. We can appreciate the small probability of a nuclear interaction (see Sec. 5-4) by noting that in 400,000 alpha-particle tracks only eight reactions of the type shown in Fig. 5-7 were found.

FIGURE 5-7   Nuclear reaction $N^{14}(\alpha,p)O^{17}$ in a cloud chamber. A thorium active deposit, not shown in the photograph, provided alpha-particle groups of 6.1 and 8.8 Mev. One alpha particle of the higher-energy group, with a computed energy of 3.9 Mev at the place of collision, produced the reaction shown. [By permission from P. M. S. Blackett and D. S. Lees, *Proc. Roy. Soc.* (*London*) **A136:** 325 (1932). Reproduced from W. Gentner, H. Maier-Leibnitz, and W. Bothe, "An Atlas of Typical Expansion Chamber Photographs," Pergamon Press, London, 1954.]

5-2b Other conservation laws.   Nuclear reactions are most conveniently discussed in the c.m. system. Conservation of angular momentum in the reaction $X(a,b)Y$ then requires

$$\mathbf{I}_a + \mathbf{I}_X + \mathit{l}_{a,X} = \mathbf{I}_b + \mathbf{I}_Y + \mathit{l}_{b,Y} \qquad (5\text{-}22)$$

where $\mathbf{I}$ is the total angular momentum of each nucleus (in units of $\hbar$), and l is the orbital angular momentum of each pair of particles about the center of mass.[1] Parity conservation requires

$$\pi_a \pi_X (-1)^{l_{a,X}} = \pi_b \pi_Y (-1)^{l_{b,Y}} \tag{5-23}$$

where π is the parity of each nuclear state involved in the reaction. These conservation laws impose restrictions on the reaction probability. But even if the conservation laws allow a reaction to proceed, the reaction rate sometimes may be so minute that its occurrence cannot be detected with available equipment.

5-3 TYPES OF NUCLEAR REACTIONS

Depending on the circumstances, it is convenient to classify nuclear reactions by the type of bombarding particle, bombarding energy, target, or reaction product. In the first case we distinguish:

> *Charged-particle reactions*, produced by p, d, α, C^{12}, O^{16} . . .
> (p = proton, d = deuteron, α = alpha particle; the last two reactions are called heavy-ion reactions)
>
> *Neutron reactions*
>
> *Photonuclear reactions*, produced by gamma rays
>
> *Electron-induced reactions*

If the bombarding energy is specified we speak informally of

> *Thermal energies* $\approx \frac{1}{40}$ ev
>
> *Epithermal energies* ≈ 1 ev
>
> *Slow-neutron energies* ≈ 1 kev
>
> *Fast-neutron energies* $\approx 0.1 - 10$ Mev
>
> *Low-energy charged particles* $\approx 0.1 - 10$ Mev
>
> *High energies* $\approx 10 - 100$ Mev

Targets are often called

> *Light nuclei*, if $A \leq 40$
>
> *Medium-weight nuclei*, if $40 < A < 150$
>
> *Heavy nuclei*, if $A \geq 150$

If the light reaction product is identical to the incident particle and has identical energy (in the c.m. system), the reaction is called elastic scattering. If only the c.m. energy is different, inelastic scattering occurs. If only gamma rays are emitted, we speak of a capture reaction. If the product nuclei have comparable masses, the reaction is called spallation or fission.

---

[1] Classically, for a two-particle system, the total orbital angular momentum about the c.m. is equal to $M_0 \mathbf{v} \times \mathbf{r}$, where $M_0 = M_1 M_2 / (M_1 + M_2)$ is the reduced mass, $\mathbf{v}$ the relative velocity of the particles and $\mathbf{r}$ the relative position vector of one particle with respect to the other. The corresponding Schrödinger equation is Eq. (2-71) or Eq. (2-146) with m_0 set equal to the reduced mass (Sec. 2.2e).

As an illustration, we give the following examples in the shorthand notation (5-3):

$N^{14}(p,p)N^{14}$	proton elastic scattering
$N^{14}(p,p')N^{14*}$	proton inelastic scattering[1]
$N^{14}(p,\alpha)C^{12}$ or C^{12*}	(p,α) reaction
$N^{14}(p,\gamma)O^{15}$ or O^{15*}	proton-capture reaction
$N^{14}(\gamma,p)C^{13}$ or C^{13*}	photonuclear reaction
$N^{14}(n,Li^6)Be^9$ or Be^{9*}	spallation reaction
$Be^9(Li^6,n)N^{14}$ or N^{14*}	heavy-ion reaction

If the reaction mechanism is clear from the experimental information, this can also be specified. We distinguish *direct* reactions and *compound-nucleus reactions* (see Fig. 5-1). Under certain circumstances, charged particles can excite the target nucleus through the electric field pulse created at the nucleus when they pass close by without penetrating the "nuclear radius." This is called *coulomb excitation*.

The variety of nuclear reactions which can occur is summarized in Table 5-1. (There are slight variances with the nomenclature given above.) Although the table may appear complex, there are many common features among nuclear reactions, which we will discuss below.

5-4   CROSS SECTIONS

The probability of occurrence of a nuclear reaction is conveniently expressed in terms of the concept of cross section.

5-4a Definition of cross section.   Since interactions in a reaction take place with individual target nuclei independently of each other, it is useful to refer the probability of a nuclear reaction to one target nucleus. Assume that in a given experiment a thin slab[2] of target material is struck by a monoenergetic beam consisting of I particles per unit time distributed uniformly over an area A, as shown in Fig. 5-8a. If the nuclear reaction produces N light product particles per unit time, we can pretend that with each target nucleus there is associated an area σ (perpendicular to the incident beam) such that if the center of a bombarding particle strikes inside of σ, there is a *hit* and a reaction is produced; and if the center of the bombarding particle *misses* σ, no reaction is produced. The quantity σ is called cross section and gives a measure of the reaction probability per target nucleus. It is a fictitious area, which need not be related to the cross sectional area (πR^2) of the struck nucleus. We could also describe the reaction probability by the ratio N/I, but this quantity depends on

---

[1] As before, an excited nucleus is denoted by a superscript asterisk. A prime on the light product particle indicates inelastic scattering.
[2] The slab should be so thin that a given bombarding particle strikes no more than one target nucleus.

TABLE 5-1   Nuclear reactions with intermediate and heavy nuclei§

This table lists the nuclear reactions occurring in each group. The symbols listed refer to the *emerging particle* in a reaction characterized by the type of target, the type of incident particle (columns), and the energy range (rows). The order of symbols in each group corresponds roughly to the order of the yields of the corresponding reactions. Reactions whose yield is usually less than about 10^{-2} of the leading one are omitted.

Abbreviations: el = elastic, inel = inelastic, res = resonances, c = coulomb excitation. The abbreviation (res) refers to all reactions listed in the box. The elastic scattering of charged particles is omitted, since it cannot easily be separated from the nonnuclear coulomb scattering. Fission is also omitted, since it occurs only with a few of the heaviest elements.

Incident Particle	Intermediate Nuclei				Heavy Nuclei			
	n	*p*	*α*	*d*	*n*	*p*	*α*	*d*
Energy of Incident Particle								
Low 0 − 1 kev	n(el) γ (res)	(N)†	(N)	(N)	γ n(el) (res)	(N)	(N)	(N)
Intermediate 1–500 kev	n(el) γ (res)	n γ α c (res)	n γ p c (res)	p n c	n(el) γ (res)	(S)‡	(S)	(S)
High 0.5–10 Mev	n(el) n(inel) p α (res for lower energies)	n p(inel) α c (res for lower energies)	n p α(inel) c (res for lower energies)	p n pn 2n c	n(el) n(inel) p γ	n p(inel) γ c	n p γ c	p n pn 2n c
Very high 10–50 Mev	2n n(inel) n(el) p np 2p α three or more particles	2n n p(inel) np 2p α three or more particles	2n n p np 2p α(inel) three or more particles	p 2n pn 3n d(inel) tritons three or more particles	2n n(inel) n(el) p pn 2p α three or more particles	2n n p(inel) np 2p α three or more particles	2n n p np 2p α(inel) three or more particles	p 2n np 3n d(inel) tritons three or more particles

† (N) = No appreciable reaction probability.
‡ (S) = Very small reaction probability.
§ By permission from Blatt and Weisskopf, 1952, chap. 9.

FIGURE 5-8   Basic experimental arrangement to determine the cross section of a nuclear reaction.   (*a*) Side view.   (*b*) View along beam direction.

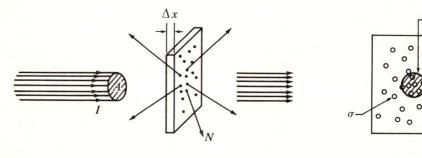

(*a*)   (*b*)

the target density as well as its thickness Δx, whereas σ is associated with an individual target nucleus.

The probability that any one bombarding particle has a hit is equal to N/I and is also equal to the projected total cross section of all target nuclei lying within the area A, as seen along the beam direction[1] (Fig. 5-8b), divided by A. If there are n target nuclei per unit volume in the target material, $nA\Delta x$, such nuclei are within reach of any bombarding particle in the beam. Each target nucleus has an associated cross section σ so that

$$\frac{N}{I} = \frac{nA\,\Delta x\,\sigma}{A} \qquad (5\text{-}24)$$

This relation can be used in two ways. First, it can serve as a definition of cross section, by writing

$$\sigma = \frac{N}{(I/A)(nA\,\Delta x)} \qquad (5\text{-}25)$$

= number of light product particles per unit time, per unit incident flux, and per target nucleus.[2]

The unit of cross section is cm² or barn $(1\text{b} = 10^{-24}\ \text{cm}^2)$. In theoretical calculations[3] Δx is usually chosen such that $nA\,\Delta x = 1$ and the flux of particles is written as

$$\frac{I}{A} = n_a v_a \qquad (5\text{-}26)$$

where $\quad n_a$ = number of bombarding particles per unit volume in the beam

v_a = relative velocity between bombarding particles and target nuclei

Second, relation (5-24) can be used to compute the yield N of light reaction products if σ is known

$$N = n\sigma\,\Delta x\,I \qquad (5\text{-}27)$$

This assumes the slab is so thin that no appreciable depletion of the beam takes place. If the slab is too thick for this assumption to be valid, since every reaction depletes the beam by one particle, we find (for a thickness dx)

$$dN = -dI$$
$$= n\sigma\,dx\,I \qquad (5\text{-}28)$$

---

[1] The target is assumed to be so thin that no nuclear cross section σ is shadowed by another target nucleus.
[2] Flux is defined here as number of particles per unit time per unit area of cross section perpendicular to the beam, i.e., I/A. See also second footnote in Sec. 2-2g.
[3] See Sec. A-2.

Integrating over the full thickness t of the slab (compare Fig. 3-14),

$$I_t = I_0 e^{-n\sigma t} \tag{5-29}$$

which is mathematically identical to Eq. (3-34). The quantity $n\sigma$ is, therefore, the linear attenuation coefficient of the beam. The ratio I_t/I_0 is sometimes called the *transmission* of the slab.

For order-of-magnitude orientation let us assume that $\sigma \approx 0.1$ b and that the interaction takes place in a cloud chamber at atmospheric pressure ($n \approx 3 \times 10^{19}$ cm^{-3}, $\Delta x \approx 10$ cm). From Eq. (5-27), we find for the probability of interaction per beam particle

$$\frac{N}{I} \approx 3 \times 10^{19} \times 0.1 \times 10^{-24} \times 10$$

$$\approx 3 \times 10^{-5}$$

This is of the order of magnitude of the experimental reaction probability which we mentioned at the end of Sec. 5-2a in connection with the $N^{14}(\alpha,p)O^{17}$ reaction (Fig. 5-7). In a solid, typical atomic densities [Eq. (3-20)] are[1] $n \approx 5 \times 10^{22}$ cm^{-3}.

In general, a given bombarding particle and target can react in a variety of ways (see example at the end of Sec. 5-3) producing a variety of light reaction products $N_1, N_2, N_3 \ldots$ per unit time. The *total cross section* is then defined in analogy with Eq. (5-25) as

$$\sigma_{\text{tot}} = \frac{N_1 + N_2 + N_3 \ldots}{(I/A)(nA\,\Delta x)} \tag{5-30}$$

It is convenient to define also a *partial cross section* for the ith process by

$$\sigma_i = \frac{N_i}{(I/A)(nA\,\Delta x)} \tag{5-31}$$

so that

$$\sigma_{\text{tot}} = \sum_i \sigma_i \tag{5-32}$$

If the partial cross sections are known, Eq. (5-31) can be rewritten in the form (5-27) to compute the rate at which the particular reaction products appear. For a thick slab, an equation similar to (3-38) must be used in order to compute this rate.

$$\frac{N_i}{I_0} = \frac{\sigma_i}{\sigma_{\text{tot}}} (1 - e^{-n\sigma_{\text{tot}}t}) \tag{5-33}$$

In many nuclear reactions, the light product particles are not produced in an isotropic manner with respect to the incident beam direction. We therefore define a *differential cross section* $d\sigma/d\Omega$ in terms of the number of light reaction products dN emitted per unit time in a small solid angle $d\Omega$ at some angle θ

[1] See Appendix B.

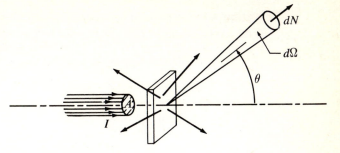

FIGURE 5-9 Basic experimental arrangement to determine a differential reaction cross section. The detector for the light reaction products subtends a small solid angle $d\Omega$ at the target.

with respect to the beam (Fig. 5-9). From Eq. (5-24)

$$\frac{1}{I}\frac{dN}{d\Omega} = \frac{nA\,\Delta x\,d\sigma/d\Omega}{A} \qquad (5\text{-}34)$$

so that differential cross section (per target nucleus) is given by

$$\frac{d\sigma}{d\Omega} = \frac{dN/d\Omega}{(I/A)(nA\,\Delta x)} \qquad (5\text{-}35)$$

In order to distinguish σ from $d\sigma/d\Omega$, a cross-section σ is sometimes called an *integrated cross section*, since

$$\sigma = \int_{\text{all space}} \frac{d\sigma}{d\Omega}\,d\Omega \qquad (5\text{-}36)$$

5-4b Energy and angular dependence of experimental cross sections.   It is the aim of any theory of nuclear reactions to explain the energy and angular dependence of cross sections in terms of certain nuclear parameters. Most cross sections have a typical behavior, shown schematically for neutrons in Fig. 5-10 and for protons in Fig. 5-11. (The target is assumed to be a medium weight nuclide.) Some actual cross sections are shown in the following sections. Unfortunately, no complete set of cross sections exists for any one nucleus.

5-4c Coulomb cross section.   The elastic scattering of low-energy charged particles is determined purely by coulomb forces. Since it is possible to compute the differential cross section by means of classical concepts this can serve as an application of Eq. (5-35). The integrated elastic scattering cross section is theoretically infinite and hence σ_{tot} has no meaning for charged particles.

Assume a beam of I charged particles per unit time bombards a thin foil. Consider the effect of a single target nucleus, as shown in Fig. 5-12a. We can show in a simple way that a particle whose impact parameter is y will be deflected

FIGURE 5-10 Schematic neutron cross sections for a medium weight nucleus. (*a*) Total cross section. (*b*) Elastic-scattering cross section. (*c*) Inelastic-scattering cross section. (*d*) Typical reaction [(*n*,α)] cross section. (*e*) Capture cross section. (*f*) Differential elastic scattering cross section.

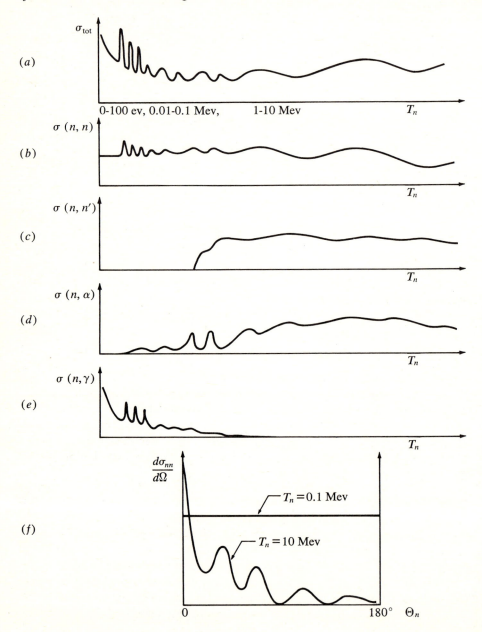

FIGURE 5-11 Schematic proton cross sections for a medium weight nucleus. (*a*) Inelastic-scattering cross section. (*b*) Typical endoergic reaction [(*p,n*)] cross section. (*c*) Typical exoergic reaction cross section. (*d*) Capture cross section. (*e*) Differential elastic-scattering cross section. For charged particles the concept of total cross section is not meaningful because the integrated elastic-scattering cross section is theoretically infinite (see Sec. 5-4c).

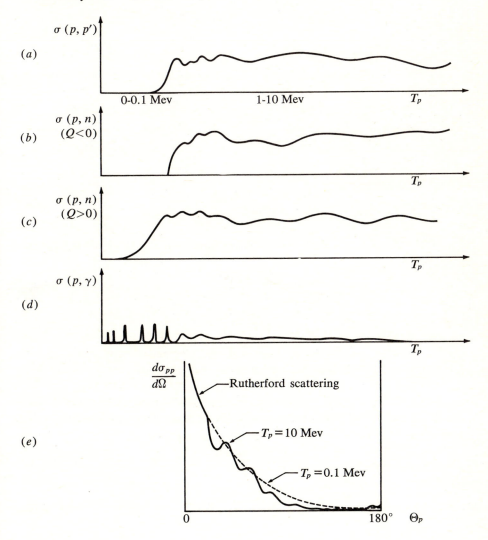

FIGURE 5-12   Effect of a single target nucleus of charge Ze on a beam of charged particles of charge ze.   (a) A particle with impact parameter y is deflected through an angle Θ. ψ is the angle between the bisector of the initial and final directions of $\mathbf{V}$ and the instantaneous radius vector $\mathbf{r}$ of the particle.   (b) Momentum diagram.

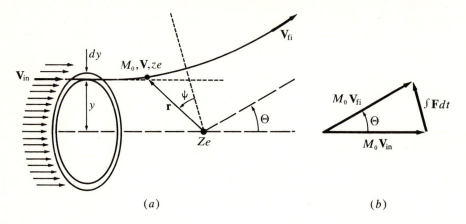

(a)   (b)

through a c.m. angle Θ given by

$$\tan \tfrac{1}{2}\Theta = \frac{D}{2y} \tag{5-37}$$

where D is the classical distance of closest approach Eq. (1-4) for a head-on collision.

$$D = \frac{zZe^2}{T_0} \tag{5-38}$$

for a particle of c.m. kinetic energy $T_0 = \tfrac{1}{2}M_0V_0^2$. This equation takes into account recoil effects if M_0 is set equal to the reduced mass [compare Eq. (4-79)] and V_0 is the relative velocity between a beam particle and the nucleus when they are far separated.

Equation (5-37) is immediately derived if we realize that the overall momentum change of the particle is caused by the impulse of the coulomb force (Fig. 5-12b). The impulse can be evaluated with the help of the law of conservation of angular momentum about the center of mass. The coulomb force is a central force and, hence, angular momentum about the force center is conserved during the collision.

$$M_0V_0y = M_0r^2\frac{d\psi}{dt} \tag{5-39}$$

where ψ is the angle between the bisector of the initial and final directions of $\mathbf{V}$ and the radius vector $\mathbf{r}$ of the particle. Now, taking all components along the bisector, the momentum diagram of Fig. 5-12b gives (since $V_{in} = V_{fi} = V_0$)

$$2M_0V_0 \sin \tfrac{1}{2}\Theta = \int F \cos \psi \, dt$$

$$= \int \frac{zZe^2}{r^2} \cos \psi \, \frac{r^2 \, d\psi}{V_0 y}$$

$$= \frac{zZe^2}{V_0 y} \int \cos \psi \, d\psi \qquad (5\text{-}40)$$

where Eq. (5-39) has been used to eliminate dt. Integrating the last expression from $-\tfrac{1}{2}(\pi - \Theta)$ to $+\tfrac{1}{2}(\pi - \Theta)$, Eq. (5-37) is obtained.

i.e. along $\int \vec{F} dt$

FIGURE 5-13   Solid-angle element for coulomb cross-section calculation.

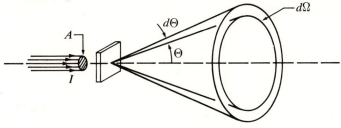

To compute the cross section, we need to find the number of beam particles per unit time scattered into a small solid angle $d\Omega$ when a beam of I particles per unit time, spread over an area A, strikes the target (see Fig. 5-13). From Fig. 5-12a, we can see that all beam particles arriving with impact parameters between y and $y + dy$ will be scattered into the solid angle[1]

$$d\Omega = 2\pi \sin \Theta \, |d\Theta| \qquad (5\text{-}41)$$

shown in Fig. 5-13. The fractional number of beam particles so scattered is $2\pi y \, dy/A$, yielding for $dN/d\Omega$

$$\frac{dN}{d\Omega} = \frac{I \, 2\pi y \, dy/A}{2\pi \sin \Theta \, |d\Theta|} \qquad (5\text{-}42)$$

Substituting into Eq. (5-35), and noting that only a single target nucleus is being considered ($nA \, \Delta x = 1$)

$$\frac{d\sigma}{d\Omega} = \frac{y \, dy}{\sin \Theta \, |d\Theta|} \qquad (5\text{-}43)$$

From Eq. (5-37)

$$dy = \frac{D \, |d\Theta|}{4 \sin^2 \tfrac{1}{2}\Theta} \qquad (5\text{-}44)$$

[1] As y increased, Θ decreased. Therefore, for a positive dy, $d\Theta$ is negative.

Writing $\sin \Theta = 2 \sin \frac{1}{2}\Theta \cos \frac{1}{2}\Theta$, we obtain

$$\frac{d\sigma}{d\Omega} = \frac{D^2}{16 \sin^4 \frac{1}{2}\Theta}$$

(5-45)

This is called the Rutherford or coulomb cross section. For order of magnitude orientation we note that, for example, 5.2-Mev protons scattered by $_{27}\text{Co}^{59}$ (see Fig. 5-14) give $D \approx 7.6$ F, and $D^2/16 \approx 0.036$ b.

We mentioned in Sec. 1-2b that if the closest distance of approach becomes

FIGURE 5-14 Differential elastic proton scattering cross section for Co^{59} divided by the Rutherford cross section, versus c.m. proton scattering angle. (By permission from F. K. McGowan, W. T. Milner, and H. J. Kim, "Nuclear Cross Sections for Charged-Particle Induced Reactions," vol. ORNL-CPX-1, Charged Particle Cross-Section Center, Oak Ridge National Laboratory, Oak Ridge, Tennessee, 1964.)

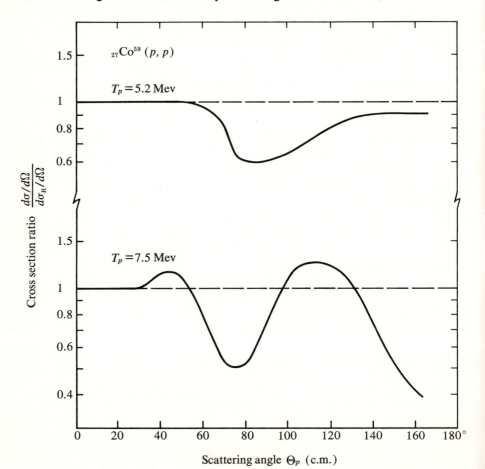

of the order of the nuclear radius R, Eq. (5-45) breaks down. For a general collision (Fig. 5-12a), this distance r_{min} depends on the scattering angle Θ and is equal to[1]

$$r_{min} = \frac{D}{2}\left(\frac{1}{\sin \frac{1}{2}\Theta} + 1\right) \tag{5-46}$$

From Fig. 5-14 we see that for protons on Co^{59}, the Rutherford scattering law breaks down at $\Theta \approx 50°$ for $T_p = 5.2$ Mev and at $\Theta \approx 30°$ for $T_p = 7.5$ Mev. Substituting into Eq. (5-46), both of these points give a consistent value of $r_{min} \approx 13$ F, a value which is considerably larger than the sum of the "Co^{59} radius" (≈ 5.5 F) and the "proton radius" (≈ 1.4 F) computed from Eq. (1-5). This is another reminder that we must not think of a sharp cutoff of the nuclear force at the nuclear surface (see Fig. 1-1).

When Eq. (5-45) is integrated over all space, an infinite result is obtained, because according to Eq. (5-37), any impact parameter, however large, gives a small deflection to a charged particle. In principle, therefore, *no* beam particle is unaffected by a target nucleus independently of its impact parameter, and the integrated cross section is infinite.[2]

At very small angles, Rutherford scattering always dominates the differential elastic-scattering cross section of charged particles. It is therefore convenient to divide the measured cross section by expression (5-45), as has been done in Fig. 5-14.

5-4d Qualitative discussion of neutron cross sections.   Before details of a quantum mechanical theory of cross sections are considered, it is useful to discuss the qualitative features that are present. In order to avoid complications due to coulomb effects, neutron cross sections will be considered.

A beam of neutrons must be represented by a (traveling) wave function of the form (2-32). For travel along the $+x$ direction, this is

$$\psi(\text{incident beam}) = ae^{ikx} \tag{5-47}$$

The most important feature of this function is the wave number k or the reduced de Broglie wavelength $\lambdabar = \lambda/(2\pi) = 1/k$. Its value, given by Eq. (2-11), is for neutrons (or protons)

$$\lambdabar \text{ (in F)} = \frac{4.55}{[T \text{ (in Mev)}]^{\frac{1}{2}}} \tag{5-48}$$

Table 5-2 evaluates this expression for some typical energies.

---

[1] R. D. Evans, 1955, app. B, sec. 3g. Note that D is the distance of closest approach for a head-on collision ($\Theta = 180°$), r_{min} for a general collision.
[2] In a real substance, the atomic electrons will shield the nuclear charge for impact parameters y larger than about 10^{-8} cm. This corresponds to such small deflection angles [see Eq. (5-37)] that the scattered particles are still within the original beam. Therefore, in most situations the integrated cross section cannot be determined.

TABLE 5-2 Reduced de Broglie wavelengths of neutrons

T	$\lambda(F)$
1 ev	4550
100 ev	455
10 kev	45.5
1 Mev	4.55
100 Mev	0.443†

† Relativistic value.

Nuclear radii [Eq. (1-5)] of medium weight nuclei are between 5 and 8 F, so that $\lambda > R$ until energies in excess of about 1 Mev are reached. We would therefore expect that the particle properties of neutrons are not too important in nuclear interactions for energies well below 1 Mev and that the wave nature of neutrons should predominate at these energies. Direct collisions (Fig. 5-1) between the incident neutron and a nucleon in the nucleus should dominate only neutron cross sections well above 1 Mev.

Let us now consider various idealized encounters of a neutron with a nucleus. If a beam of neutrons strikes a perfectly reflecting nucleus, only elastic scattering can occur. The neutron wave is reflected and diffracted by the nucleus as indicated schematically in Fig. 5-15a. These two waves interfere. The

FIGURE 5-15 Effect of a nucleus on a neutron wave (schematic). (*a*) Perfectly reflecting nucleus. Reflection and diffraction occur. (*b*) Perfectly absorbing nucleus. Only diffraction occurs. (*c*) Partially transparent nucleus. The wave transmitted through the nucleus can interfere with the reflected and diffracted waves.

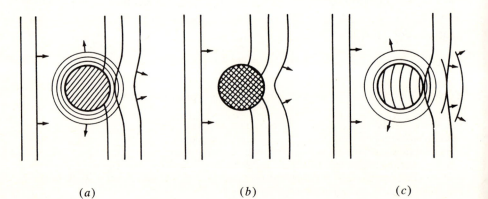

(*a*)   (*b*)   (*c*)

(integrated) elastic scattering cross section at low energies ($\lambda \gg R$) is found[1] to be equal to $4\pi R^2$.

If the nucleus is perfectly absorbing, there will be no reflected wave, but diffraction scattering still occurs (Fig. 5-15b). The high-energy elastic-scattering cross section has the approximate value πR^2, and the total cross section is approximately equal to[2] $2\pi R^2$. If the nucleus is partially transparent to the incident neutrons, the transmitted wave interferes with the reflected and diffracted waves (Fig. 5-15c). For certain energies, i.e., wavelengths, we may expect that constructive interference occurs, for others, destructive interference. Resonances in cross sections are caused by such interference phenomena. Where resonances occur, cross sections are of the order of $\pi \lambda^2$, rather than $4\pi R^2$, which holds between resonances; hence, slow neutron cross sections can be much greater than geometrical cross sections.

FIGURE 5-16 Classical interpretation of orbital angular momentum imparted in a nuclear reaction. M_0 is the reduced mass $M_a M_X/(M_a + M_X)$.

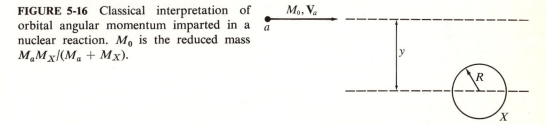

In a more detailed discussion of cross sections, it is useful to consider the orbital angular momentum brought into the system (bombarding particle + target nucleus) by the reaction because this may influence the cross section as a result of the selection rules (5-22) and (5-23). Classically, if the bombarding particle has an impact parameter y with respect to the nucleus (Fig. 5-16), the orbital c.m. angular momentum is equal to $M_0 V_a y$ [see Eq. (5-39) and footnote after Eq. (5-22)], where V_a is the relative velocity of the two particles when they are far separated. In reality, though, the orbital angular momentum is quantized and equal to $l_{a,\,X}\hbar$. (This is abbreviated $l_a\hbar$ below.) We can therefore identify roughly[3]

$$l_a\hbar \approx M_0 V_a y \qquad (5\text{-}49)$$

$$l_a \approx \frac{y M_0 V_a}{\hbar}$$

$$\approx \frac{y}{\lambda} \qquad (5\text{-}50)$$

where λ is the c.m. reduced de Broglie wavelength of the neutron.

[1] Burcham, 1964, pp. 518–520.
[2] See Sec. A-2, Eq. (A-27).
[3] Such arguments can be shown to be more correct for large quantum numbers than for small ones, in accordance with Bohr's correspondence principle. Compare the discussion following Eq. (2-89).

If $y > R$ (Fig. 5-16), the bombarding particle should not have much effect on the target nucleus because it would be outside the range of nuclear forces. Therefore, we expect the most important nuclear interactions to occur with those bombarding particles which have orbital angular momenta *less than or equal to* a maximum value

$$l_a(\text{max}) \approx \frac{R}{\lambda} \tag{5-51}$$

Referring to Table 5-2, we find that interactions which occur with bombarding energies below about 0.1 Mev should be predominantly s wave, i.e., have $l_a = 0$. (For a hydrogen target this statement obtains to approximately 10-Mev neutron lab. energy.) As a result, the differential elastic-neutron-scattering cross section is isotropic in the c.m. system up to these energies.[1] This agrees with observation.

The decomposition of the incident wave into partial waves each associated with a definite orbital angular momentum is called partial wave analysis and is described briefly in Appendix A-2. More details can be found elsewhere.[2]

5-5 COMPOUND-NUCLEUS REACTIONS

For bombarding energies below 0.1 to 1 Mev, nuclear reactions generally proceed through the compound-nucleus mechanism (Fig. 5-1). The reason is that once a particle finds itself within the nucleus, the reflection coefficient at the edge of the potential well (Fig. 5-2) is close to unity.[3] For a bombarding particle of energy T_0 entering a square well of depth V_0, the reflection coefficient, given by Eq. (2-162), is approximately equal to

$$1 - 4(T_0/V_0)^{\frac{1}{2}} \quad \text{if} \quad T_0 \ll V_0 \tag{5-52}$$

If $V_0 \sim 40$ Mev, a typical value, the reflection coefficient for $T_0 = 0.1$ Mev is approximately equal to 0.8. The particle is therefore likely to stay an appreciable time within the nucleus, and the chain of processes indicated in Fig. 5-1 can proceed to the compound-nucleus stage.

The characteristic feature of experimental cross sections in this energy range is the appearance of many sharp resonances (Figs. 5-17, 5-18). As we mentioned in Sec. 5-1, the detailed nature of the resonances is not easy to describe in terms of the shell model, but consists of very complicated excitations of many nucleons in the nucleus. From a wave point of view, though, the resonances are caused by the interference between the wave emerging from the nucleus and the diffracted and reflected waves of the bombarding particle[4] (Fig. 5-15c).

[1] We used this fact in the derivation of Eq. (3-30). See also Sec. A-2.
[2] Burcham, 1963, sec. 14-2.
[3] The expression for the reflection coefficient at a potential discontinuity is independent of the direction of travel.
[4] Blatt and Weisskopf, 1952, chap. 8, sec. 7.

FIGURE 5-17 Total and elastic neutron scattering cross section of cadmium. The resonance at 0.18 ev is caused by Cd^{113} (12 percent abundance). The dashed line at low energies is for a $1/v$ dependence of the cross section. (Adapted from D. J. Hughes and J. A. Harvey, "Neutron Cross Sections," 1st ed., and D. J. Hughes and R. B. Schwartz, "Neutron Cross Sections," 2d ed., Brookhaven National Laboratory, Upton, New York, 1955 and 1958.)

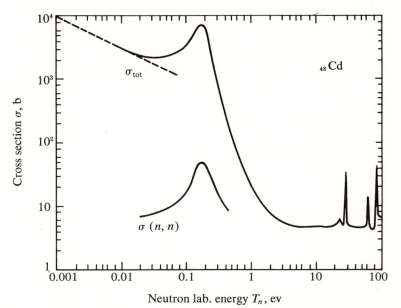

Neutron lab. energy T_n, ev

The basic assumption of the compound-nucleus model is that the compound nucleus has been formed in such a complex manner that it has "forgotten" how it has been formed. The cross section for the reaction $X(a,b)Y$ can then be split up into a formation cross section of the compound nucleus C^* corresponding to the process

$$a + X \rightarrow C^* \tag{5-53}$$

and the fractional probability that C^* breaks up into particles[1] $b + Y$. We can therefore write

$$\sigma(a,b) = \sigma_{a,C}(T_0)P_b(E) \tag{5-54}$$

where T_0 is the (c.m.) bombarding energy and E the corresponding excitation energy of the compound nucleus. These energies are related as shown in Fig. 5-19.

---

[1] A particular breakup mode is called a *channel*. For example, $a + X$, $a' + X^*$, $b + Y$ are different channels.

5-5a Formation of the compound nucleus.   For the present, we will assume that a is a spinless, neutral particle, so that spin and coulomb effects can be ignored. If the bombarding energy T_0 is such that C is formed in an excited state of energy E^*, this state will of course be a virtual state, because it can always decay back to $a + X$. Therefore, it will have a finite width Γ [Eq. (4-32)], due to its finite

FIGURE 5-18   Total neutron cross section of sulfur (95% S^{32}). (By permission from D. J. Hughes and J. A. Harvey, "Neutron Cross Sections," 1st ed., and D. J. Hughes and R. B. Schwartz, "Neutron Cross Sections," 2d ed., Brookhaven National Laboratory, Upton, New York, 1955 and 1958.)

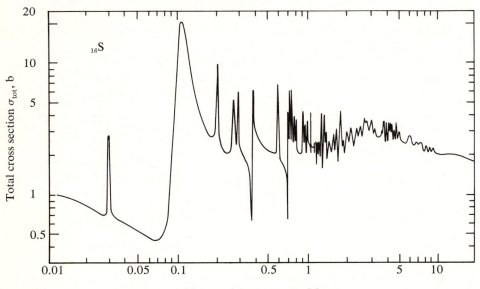

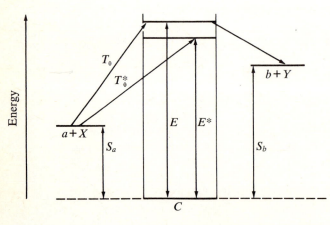

FIGURE 5-19   Reaction $X(a,b)Y$ proceeding through a compound nucleus. This figure should be compared with Fig. 5-6. S_a and S_b are the separation energies of a and b, respectively, from C. T_0 is the c.m. energy of a [$T_0 = T_a M_X / (M_a + M_X)$].

lifetime. The cross section for formation of the compound nucleus is expected to be proportional to the probability of finding the nucleus at an energy E, given by Eq. (4-41). A complete quantum mechanical calculation[1] shows that

$$\sigma_{a,C} = \pi\lambda^2 \frac{\Gamma_a \Gamma}{(E - E^*)^2 + \Gamma^2/4} \tag{5-55}$$

where $\Gamma_a/\hbar$ is the decay constant for decay of the compound state into channel $a + X$. We call Γ the total width of the state and Γ_a the partial width for decay into $a + X$. In general

$$\Gamma = \Gamma_a + \Gamma_b + \Gamma_{b'} + \Gamma_{b''} + \cdots \tag{5-56}$$

where $\Gamma_b \cdots$ are the partial widths for any other channels energetically allowed. This equation follows immediately from the relation between the total and partial decay constants [compare Eq. (4-12)].

The wavelength λ in Eq. (5-55) is the reduced de Broglie wavelength of a in the c.m. system. We note that for neutrons (or protons)

$$\pi\lambda^2 \text{ (in barns)} = \frac{0.65}{T_0 \text{ (in Mev)}} \tag{5-57}$$

It is convenient to express the resonance energy E^* in terms of the corresponding (c.m.) bombarding energy T_0^* (see Fig. 5-19) so that Eq. (5-55) becomes

$$\sigma_{a,C} = \pi\lambda^2 \frac{\Gamma_a \Gamma}{(T_0 - T_0^*)^2 + \Gamma^2/4} \tag{5-58}$$

Many tabulations give the lab. energies T_a and T_a^* corresponding to T_0 and T_0^*. If these energies are used in Eq. (5-58), the widths must also be converted to the lab. system by using a relation similar to (5-16)

$$\Gamma_i(\text{lab.}) = \Gamma_i(\text{c.m.})(M_a + M_X)/M_X \tag{5-59}$$

If spin is considered, the right-hand side of Eq. (5-58) must be multiplied by the factor

$$g_J = \frac{2J + 1}{(2I_a + 1)(2I_X + 1)} \tag{5-60}$$

where J is the total angular momentum of the compound state and the other spins have been defined before [Eq. (5-22)]. A given compound state can be formed only by those orbital angular momenta l_a which satisfy the relations

$$\mathbf{I}_a + \mathbf{I}_X + \mathbf{l}_a = \mathbf{J} \tag{5-61}$$

$$\pi_a \pi_X (-1)^{l_a} = \pi_J \tag{5-62}$$

---

[1] Blatt and Weisskopf, 1952, pp. 398ff. For a derivation starting with Eq. (4-41) see Burcham, 1963, p. 532.

where π_J is the parity of the compound state. In addition, conditions (5-22) and (5-23) must be satisfied.

5-5b Decay of the compound nucleus.   From the definition (5-56) it follows that the probability P_b of decay of a compound state [Eq. (5-54)] into channel $b + Y$ is given by

$$P_b = \frac{\Gamma_b}{\Gamma} \tag{5-63}$$

where Γ_b is the appropriate partial width. Combining this with Eqs. (5-58) and (5-60) we find

$$\sigma(a,b) = g_J \pi \lambda^2 \frac{\Gamma_a \Gamma_b}{(T_0 - T_0^*)^2 + \Gamma^2/4} \tag{5-64}$$

This is known as the *Breit-Wigner resonance formula*. Equation (5-64) is valid for all channels except the elastic scattering channel for which the interference with diffraction and reflection scattering have to be considered. For neutrons one finds

$$\sigma(n,n) = 4\pi\lambda^2 \left[g_J \left| \frac{\Gamma_n}{2(T_0 - T_0^*) + i\Gamma} + e^{i\phi_{l_n}} \sin \phi_{l_n} \right|^2 + (1 - g_J) \sin^2 \phi_{l_n} \right] \tag{5-65}$$

where ϕ_{l_n} is an energy-dependent quantity known as the *hard-sphere phase shift*. For $l_n = 0$, $\phi = R/\lambda$.

We will see [Eq. (5-69)] that as $T_0 \rightarrow 0$, $\Gamma_n \rightarrow 0$. Since, under the same conditions, $\sin \phi \rightarrow R/\lambda$, the elastic-scattering cross section approaches $4\pi R^2$ at very low bombarding energies. It is usual in low-energy neutron physics to express the elastic-scattering cross section in terms of a so-called *scattering length*[1] a as

$$\sigma(n,n)_{T_0 \rightarrow 0} = 4\pi a^2 \tag{5-66}$$

In the simple case just discussed, $|a| = R$; but in general this is not true, either because $l_n \neq 0$ or because the influence of resonances is not negligible (see Sec. 5-6a).

Figure 5-20 shows some typical cross-section shapes calculated with Eqs. (5-64) and (5-65). Both shapes can be recognized in Figs. 5-17, 5-18.

Before discussing special applications of the Breit-Wigner formula we make a few remarks about partial widths. Any partial width Γ_b is simply another way of representing the decay constant λ_b for the process

$$C^* \rightarrow b + Y \tag{5-67}$$

This is eactly like radioactive (alpha or gamma) decay and so we can apply the concepts embodied in Eqs. (4-69), (4-95), or (4-123). From the latter we find

[1] See Sec. A-2.

FIGURE 5-20   Typical Breit-Wigner resonance shapes.   (*a*) Reaction cross section. (*b*) Elastic scattering cross section for *s*-wave neutrons.

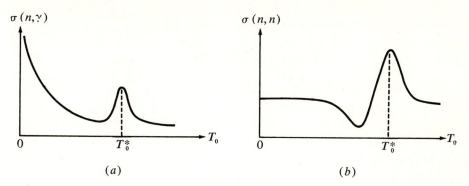

that the width must be proportional to the density of final states (4-129) which for *particles* is proportional to

$$p^2 \frac{dp}{dE} = pM_{0b} \sim v_b \tag{5-68}$$

the relative velocity of the particles in the channel $b + Y$. Here M_{0b} is the reduced mass of channel $b + Y$. From Eq. (4-95) we realize that for *charged* particles the width must also be proportional to a coulomb penetrability factor. But even in the absence of a coulomb barrier, the centrifugal barrier alone (Fig. 4-16*b*) can decrease the width. To take both these factors into account one usually writes

$$\Gamma_b = 2k_b R P(b, Y)\gamma_b{}^2 \tag{5-69}$$

where the factor 2 is introduced for convenience, and

$$k_b = \text{c.m. wave number of channel } b + Y = M_{0b}v_b/\hbar$$
$$R = \text{``nuclear radius''} \approx R_0(A_b{}^{1/3} + A_Y{}^{1/3}) \ (A = \text{mass number})$$
$$P(b, Y) = \text{penetration factor, which is unity for } s\text{-wave neutrons}$$
$$\gamma_b{}^2 = \text{experimentally determined constant, called the } reduced\ width$$

The largest possible particle width that any channel can have is given by the estimate

$$\Gamma_b(\text{max}) \approx \frac{\hbar}{t} \tag{5-70}$$

where t is the time necessary for particle b to pass *by* the nucleus

$$t \approx \frac{R}{v_b} \tag{5-71}$$

$$\Gamma_b(\text{max}) \approx \frac{\hbar v_b}{R} \approx k_b R\left(\frac{\hbar^2}{M_{0b}R^2}\right) \tag{5-72}$$

This has been written in the form of Eq. (5-69). The quantity in parentheses is called the *single particle width* [see Eq. (2-145)],[1] and one often compares experimental reduced widths to it. If $\gamma_b{}^2 \approx \hbar^2/(M_{0b}R^2)$, the compound state in question can be thought of as consisting mainly of particle b moving in a potential provided by Y. Values of reduced widths as small as 10^{-6} of the single particle width are common, indicating the complex nature of compound-nucleus states.

Gamma-ray widths are given directly by Eq. (4-69). Normally, particle widths (5-69) exceed gamma-ray widths if the particle kinetic energy in the particular channel is above 1 kev. For (virtual) levels of the compound nucleus, gamma decay is therefore rare unless the levels lie within a few kev of the lowest separation energy, or special circumstances inhibit particle decay.

5-5c Special cases.

1 Low-energy neutron reaction cross sections.

Neutron reactions which are energetically permitted for zero-energy neutrons (capture or exoergic reactions) have a $1/v$ cross-section dependence at low energies, if no resonance lies too close to the neutron separation energy of the compound nucleus (i.e., near zero neutron kinetic energy). Under this condition, T_0 can be neglected with respect to T_0^* in Eq. (5-64)

$$\sigma(n,b) \approx g_J \pi \lambda^2 \frac{\Gamma_n \Gamma_b}{T_0^{*2} + \Gamma^2/4} \tag{5-73}$$

If $\Gamma \ll T^*$, or if Γ does not depend sensitively on the neutron energy, then since $\lambda \sim 1/v_n$ [Eq. (5-48)] and $\Gamma_n \sim v_n$ [Eq. (5-69)]

$$\sigma(n,b) \sim \frac{1}{v_n} \tag{5-74}$$

This type of low-energy behavior can be seen in Fig. 5-17. In fact, the major part of the low-energy Cd total neutron cross section is caused by the $Cd^{113}(n,\gamma)$ reaction and hence Eq. (5-64) can be applied to the prominent resonance. A detailed fit of the (n,n) and (n,γ) cross sections below 2 ev yields the following values for the resonance parameters:

$$T_0^* = 0.18 \text{ ev}$$
$$\Gamma_n = 0.65 \times 10^{-3} \text{ ev} \qquad \text{(assuming } J = 1\text{)}$$
$$\Gamma_\gamma = 0.11 \text{ ev} \qquad \text{(assuming } J = 1\text{)}$$

Comparison of Γ_n with Eq. (5-69) gives $\gamma_n{}^2 = 0.15$ ev for the reduced width, which can be compared to a single-particle width [Eq. (5-72)] of approximately 1 Mev. Therefore, this particular state of the compound nucleus Cd^{114} has

---

[1] The width in Eq. (2-145) refers to a quasi-bound state, whereas the width estimates in Eq. (5-72) refer to a free particle. The difference lies in the time t: In Eq. (2-141), t refers to the time for passing *through* the nucleus; in Eq. (5-70), t is the time needed to pass *by* the nucleus.

a very complex structure. Experimentally one finds that only 0.1 percent of the gamma rays go to the ground state[1] of Cd^{114} so the width for that transition alone would be $\Gamma_\gamma \approx 10^{-4}$ ev. The Weisskopf single-particle estimate [Eq. (4-69)] for an $M1$ transition of 9 Mev, corresponding to the $1^+ \rightarrow 0^+$ ground state transition, is $\Gamma_\gamma \approx 15$ ev, showing again that the compound state is not at all single-particle-like.

FIGURE 5-21 Inelastic neutron scattering. (*a*) Energetics. (*b*) Typical cross section. T_0^* corresponds to a resonance of the compound nucleus.

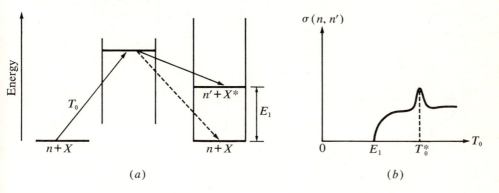

(*a*)   (*b*)

2 Inelastic neutron scattering.
For this reaction Eq. (5-64) gives

$$\sigma(n,n') = g_J \pi \lambda^2 \frac{\Gamma_n \Gamma_{n'}}{(T_0 - T_0^*)^2 + \Gamma^2/4} \tag{5-75}$$

Close to the threshold of the reaction, which may correspond to the energy E_1 of the first excited state of the target nucleus (Fig. 5-21*a*), the energy dependence of Eq. (5-75) is dominated by $\Gamma_{n'}$. If s-wave neutrons are produced so that the penetration factor in Eq. (5-69) is unity,

$$\Gamma_{n'} \sim v_{n'}$$
$$\sim (T_0 - E_1)^{\frac{1}{2}} \tag{5-76}$$

In Fig. 5-22*b* this energy dependence (schematically shown in Fig. 5-21*b*) disappears in the resonance structure.

3 Emission of charged particles.
If charged particles are emitted in an exoergic neutron reaction, the low-energy dependence of the cross section is dominated by the $1/v_n$ dependence. In an endoergic reaction, the energy dependence near the threshold is dominated

---

[1] A knowledge of slow-neutron capture gamma-ray spectra is a useful tool in nuclear level scheme investigations. See Groshev et al., 1959.

FIGURE 5-22 (*a*) Neutron reaction cross sections for $_{16}S^{32}$. (*b*) Inelastic neutron scattering cross section for $_{14}Si^{28}$. (By permission from J. R. Stehn et al., "Neutron Cross Sections," 2d ed., suppl. no. 2, vol. 1, Sigma Center, Brookhaven National Laboratory, Upton, New York, 1964.)

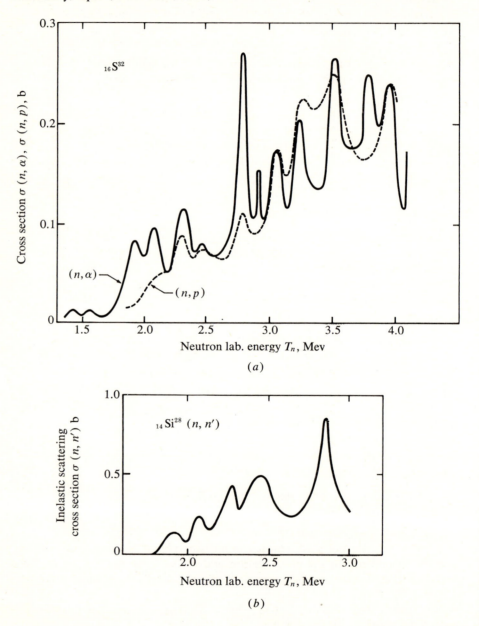

(*a*)

(*b*)

by the coulomb penetration factor (4-86) in Eq. (5-69)

$$\Gamma_b \sim e^{-\gamma} \qquad (5\text{-}77)$$

where, according to Eq. (4-94),
$$\gamma \sim \frac{1}{v_b} \qquad (5\text{-}78)$$

The cross section therefore rises very slowly above threshold, as indicated schematically in Fig. 5-10d. Experimental cross sections are shown in Fig. 5-22a.

   4 Charged-particle induced reactions.
   At low energy, all exoergic reactions of this type are dominated by the coulomb penetration factor of the *incident* particle. Endoergic charged-particle reactions have an energy dependence near threshold of the form (5-76) or (5-77) for neutron or charged-particle emission, respectively. Figure 5-23 is an example of an endoergic reaction in which neutrons are emitted.

5-6 DIRECT REACTIONS

From the general discussion in Secs. 5-1 and 5-5, it appears that with increasing energy of the bombarding particle the first stage of the interaction process shown in Fig. 5-1 becomes more important and the later stages less important. There is considerable experimental evidence that, on the whole, this is correct, although examples of reactions proceeding through very highly excited compound states are known.

5-6a Optical model.   In a theory which emphasizes the first stage of a nuclear reaction (Fig. 5-1), the interaction of the incident particle with the nucleus can be represented by a potential. If the absorption into the compound system is a relatively minor effect, it can be taken into account in a phenomenological way by adding a complex term to the effective potential

$$V_{\text{eff}} = V + iU \qquad (5\text{-}79)$$

A simple one-dimensional calculation shows that such a potential will produce absorption of a wave function.
   Consider a beam of particles of mass M_0 encountering a complex potential step as indicated in Fig. 5-24a. Outside the potential step, the form of the entering wave is ae^{ikx} [see Eq. (5-47)]. Inside the potential step, the wave function will be of the form

$$a'e^{ik'x} \qquad (5\text{-}80)$$

where k' is such that (see Fig. 5-24a)

$$\frac{\hbar^2 k'^2}{2M_0} = T + V_0 + iU_0 \qquad (5\text{-}81)$$

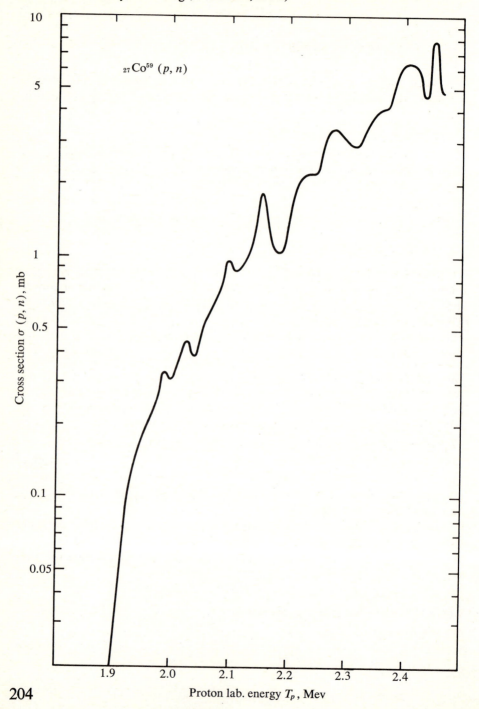

FIGURE 5-23 (p,n) cross section for Co^{59}. (By permission from F. K. McGowan, W. T. Milner, and H. J. Kim, "Nuclear Cross Sections for Charged-Particle Induced Reactions," vol. ORNL-CPX-1, Charged Particle Cross-Section Center, Oak Ridge National Laboratory, Oak Ridge, Tennessee, 1964.)

$_{27}\text{Co}^{59}$ (p, n)

Cross section σ (p, n), mb

Proton lab. energy T_p, Mev

FIGURE 5-24 The effect of a complex potential on a wave function. (*a*) Potential energy diagram. (*b*) Schematic wave function (the wave function is complex).

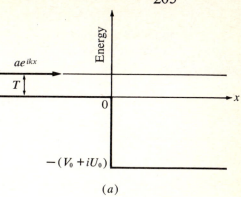

(*a*)

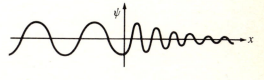

(*b*)

The incident kinetic energy is denoted by $T = \frac{1}{2}\hbar^2 k^2/M_0$. Equation (5-81) obviously requires that k' be complex. Defining its real and imaginary parts by

$$k' = K + \frac{i}{L} \tag{5-82}$$

and substituting into Eq. (5-80), the wave function inside will be

$$a'e^{-x/L}e^{iKx} \tag{5-83}$$

In other words, the wave is absorbed as it penetrates into the potential step. Substitution of Eq. (5-82) into (5-81) gives

$$L = \frac{\hbar^2 K}{U_0 M_0} \tag{5-84}$$

and

$$K \approx \left[\frac{2M_0(T + V_0)}{\hbar^2}\right]^{\frac{1}{2}} \tag{5-85}$$

if $1/L \ll K$.

This model of the nuclear interaction has been particularly successful in explaining total and elastic cross sections at high energies. From detailed fits of experimental cross sections, the values of V_0 and U_0 listed[1] in Table 5-3 have been

---

[1] The values in Table 5-3 apply to a rounded well (Sec. 2-5b) and are only meant to indicate orders of magnitude.

found. Values of K and L calculated from Eqs. (5-84) and (5-85) are also given in the table. As the incident energy increases, nuclei become more absorbent (L decreases).

TABLE 5-3 Approximate optical model parameters for protons and neutrons†

T, Mev	V_0, Mev	U_0, Mev	K, F^{-1}	L, F
$0 - 4$	50	3	1.6	22
10	50	7	1.7	10
17	50	8.5	1.8	9
40	35	15	1.9	5

† From H. Feshbach, The Complex Potential Model, in F. Ajzenberg-Selove (ed.), "Nuclear Spectroscopy," Academic Press Inc., New York, 1960, part B, Chap. 6.D, by permission.

The model predicts broad resonances in cross sections as a function of energy. One of these can be recognized in Fig. 5-18 near 3 Mev. Compound resonances are, of course, omitted from the description of the interaction in terms of expression (5-79). Nevertheless, the model predicts that some features of the first step of the interaction process should *show through* even at the compound nucleus stage. For example, whenever the bombarding particle is in a virtual state of the potential well, enhanced compound-nucleus cross sections are predicted. A virtual state requires that the incident particle form an approximate standing wave in the potential well [see Eq. (2-143)]

$$n \cdot \tfrac{1}{2}\lambda_{\text{inside}} \approx R \tag{5-86}$$

where $n = $ an integer
$\lambda_{\text{inside}} \approx 2\pi/K$ is the wavelength inside the well
$R = $ nuclear radius
Substituting the low-energy value of K from Table 5-3, we predict that enhanced cross sections should occur for nuclei with mass numbers[1]

$$A \approx 2n^3$$
$$\approx 2, 16, 54, 128, 250 \tag{5-87}$$

Such effects have been observed. For example, the low-energy elastic neutron scattering cross section, instead of having the value $4\pi R^2$ expected from Eq. (5-65), has peaks near the values of A predicted by Eq. (5-87). This is shown in Fig. 5-25. The reduced widths $\gamma_n{}^2$ are similarly affected.

[1] This relation has to be modified for the permanently deformed nuclei (Sec. 2-5d) near $A \approx 150$.

The optical model has also been successful in explaining the forward peaking of the neutron differential elastic-scattering cross section, at higher energies, of which Fig. 5-26 is a typical example. The forward peaking is caused by the interference effects discussed in connection with Fig. 5-15c.

FIGURE 5-25 The apparent nuclear radius R_{app}, inferred from the low-energy neutron scattering cross section $\sigma(n,n) = 4\pi(R_{app})^2$, as a function of the mass number A. The points are experimental. The dashed line represents $R_{app} = 1.35A^{\frac{1}{3}}$. The other curves are obtained by means of various optical-model theories. (By permission from K. K. Seth, quoted in Marion and Fowler, 1963.)

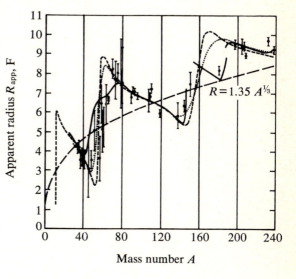

5-6b Surface interaction model.   We note from Table 5-3 that for high incident energies the mean free path $\frac{1}{2}L$ becomes comparable to the nuclear radius.[1] If we push this effect to the extreme, we can assume that at these energies all nuclear interactions occur only at the surface. Such a model would be especially appropriate for complex bombarding particles, for which the mean free path is smaller than for neutrons and protons.

We can show that the model will give characteristic diffraction patterns for elastic scattering. An oversimplified view of the surface interaction is sketched in Fig. 5-27. If only points A and B in the nucleus are assumed to rescatter the incident wave, constructive interference at a c.m. angle Θ requires that

$$CB + BD = n\lambda \tag{5-88}$$

where n is an integer and λ is the wavelength of the incident radiation. Hence, peaks in the elastic scattering cross section should occur whenever

$$2 \cdot 2R \sin \tfrac{1}{2}\Theta = n\lambda \tag{5-89}$$

---

[1] The mean free path in the nucleus is equal to $\frac{1}{2}L$, because it is defined for a flux of particles. From Eq. (5-83), the flux is proportional to $|a'|^2 e^{-2x/L}$. (See second footnote in Sec. 2-2g.)

FIGURE 5-26 Differential elastic neutron scattering cross section of Co^{59} for lab. neutron energies of 1 and 14 Mev versus the cosine of the c.m. scattering angle. (By permission from M. D. Goldberg, V. M. May, and J. R. Stehn, "Angular Distributions in Neutron-Induced Reactions," 2d ed., vol. 2, Sigma Center, Brookhaven National Laboratory, Upton, New York, 1962.)

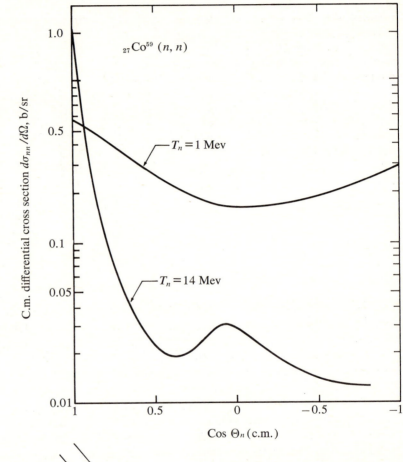

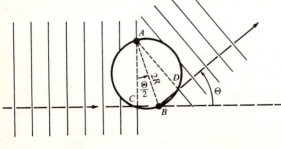

FIGURE 5-27 Oversimplified view of surface interaction model. In the figure it is assumed that only the points A and B on the surface of the nucleus scatter the incident particles into the angle Θ. All other incident particles are assumed to be completely absorbed.

FIGURE 5-28 Differential elastic scattering cross section for alpha particles on Fe^{58}, divided by the Rutherford scattering cross section, as the function of the c.m. scattering angle. The incident lab. energy is 64 Mev. (By permission from F. K. McGowan, W. T. Milner, and H. J. Kim, "Nuclear Cross Sections for Charged-Particle Induced Reactions," vol. ORNL-CPX-1, Charged Particle Cross-Section Center, Oak Ridge National Laboratory, Oak Ridge, Tennessee, 1964.)

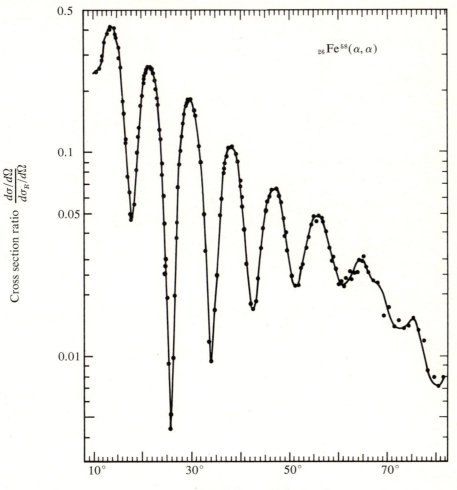

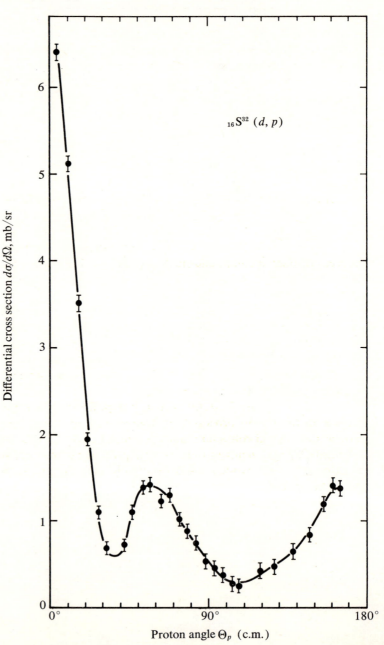

FIGURE 5-29 Differential cross section for the $S^{32}(d,p)S^{33}*$ reaction. The incident deuteron energy is 4.0 Mev. Only those protons are detected which leave S^{33} in its first excited state at 0.84 Mev. [By permission from I. B. Teplov and B. A. Iurev, *J. Exptl. Theoret. Phys. (U.S.S.R.)*, **34**: 334 (1958); English Transl. *Soviet Phys. JETP*, **7**: 233, (1958).]

$_{16}S^{32}\ (d, p)$

Differential cross section $d\sigma/d\Omega$, mb/sr

Proton angle Θ_p (c.m.)

Figures 5-14 and 5-28 show differential elastic scattering cross section for protons and alpha particles in which the diffraction peaks follow approximately relation (5-89). We recognize that this provides a means for obtaining the nuclear radius. Indeed, expression (1-5) was obtained experimentally from such experiments.

5-6c Stripping reactions.   If a complex bombarding particle encounters a nucleus, it can break up on impact such that only one part of it interacts strongly with the nucleus and the other part leaves with practically no interaction. Evidence for such processes has been found, especially for incident deuterons and other relatively loosely bound structures. It is a characteristic feature of these reactions that the noninteracting component of the bombarding particle travels off predominantly in the forward direction, i.e., in the direction of the incident beam. Figure 5-29 gives a typical example.

5-7 FISSION

A reaction $X(a,b)Y$ is called fission, if b and Y have comparable masses. Some nuclei fission spontaneously. Usually fission is produced only if sufficient energy is given to a nucleus by capture of a slow neutron or by bombardment with $n, p, d \cdots$ or gamma rays. As far as we know, the fission process always proceeds through a compound-nucleus stage. The compound nucleus breaks up into two parts with some prompt neutron emission. The reason for this will become apparent immediately.

The fission process was discovered by Hahn and Strassmann (1939) by means of radiochemical experiments. They showed that the bombardment of uranium by neutrons produces elements from the middle of the periodic table, and not trans-uranium elements, as was believed previously.

The two main nuclear components, called *fission fragments,* do not have equal masses because of energetics. The mass distribution is probably influenced by shell effects. Figure 5-30 shows that prompt fission fragments are not stable, because in the fission process both fragments keep the same neutron-proton ratio as the original compound nucleus which lies close to the stability line. The fragments are very neutron rich. Hence, prompt neutron emission is favored. Negative beta and gamma decay eventually take the fission products toward the stability line. In some cases excited states, which lie above the neutron separation energy of the particular nucleus, are populated by beta decay, so that *delayed* neutron emission occurs.[1]

By checking the line of constant N/Z for a typical fissioning nucleus like $_{92}U^{236}$ with the N-Z plot, Fig. 2-10, we see that this line passes close by the doubly

---

[1] The *delay* is with respect to the initial fission event. Neutron decay of any virtual state occurs within times of the order of the estimate (5-71), that is, 10^{-17} sec.

magic nucleus $Z = 50$, $N = 82$. We would therefore expect $A = 132$ to be a prominent mass number on the final mass-yield curve. This checks quite well with experiment (Fig. 5-31), but probably does not furnish the entire explanation for the shape of the mass-yield curve.

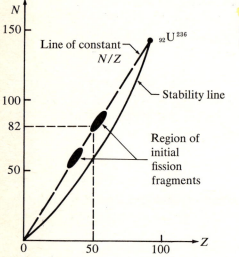

FIGURE 5-30 Location of fission fragments with respect to stability line. The example shown is for the fission of the compound nucleus $_{92}U^{236}$.

$\longrightarrow \beta^-$ decay of fragments

5-7a Energy release in fission.

The kinetic energies of the prompt fission fragments can be computed from the semiempirical mass formula. For example, consider the fission process

$$U^{235} + \text{thermal } n \rightarrow Y_1 + Y_2 \tag{5-90}$$

where Y_1 and Y_2 have the same N/Z ratio as U^{236}. Then

$$
\begin{aligned}
Q(\text{prompt}) &= T_{Y_1} + T_{Y_2} \\
&= [M(U^{235}) + M_n - (M_{Y_1} + M_{Y_2})]c^2 \\
&= B_{\text{tot}}(Y_1) + B_{\text{tot}}(Y_2) - B_{\text{tot}}(U^{235})
\end{aligned}
\tag{5-91}
$$

The total binding energies can be calculated from Eq. (2-127) with the result $Q(\text{prompt}) \approx 170$ Mev.

The overall energy release for the final fission products includes the energy release by beta rays, gamma rays, and antineutrinos, as well as the energy carried off by neutrons. The Q value then is

$$Q(\text{overall}) = B_{\text{tot}}(Y_1') + B_{\text{tot}}(Y_2') - B_{\text{tot}}(U^{235}) \tag{5-92}$$

where Y_1' and Y_2' are the final fission products which lie near the stability line. If the mass numbers of these products are 132 and 100, respectively, assuming 4 neutrons are released, Fig. 2-8 gives for a rough estimate

$$Q(\text{overall}) \approx 132 \times 8.3 + 100 \times 8.5 - 235 \times 7.5 \text{ Mev}$$

$$\approx 210 \text{ Mev}$$

A more accurate calculation from actual masses gives the values in Table 5-4. Noting that antineutrinos do not produce any usable energy, we find that 3.2×10^{10} fissions/sec produce 1 watt of power.

FIGURE 5-31 Mass-yield curve for the thermal-neutron fission of U^{233}, U^{235}, and Pu^{239}. (By permission from A. M. Weinberg and E. Wigner, "The Physical Theory of Neutron Chain Reactors," University of Chicago Press, Chicago, 1959.)

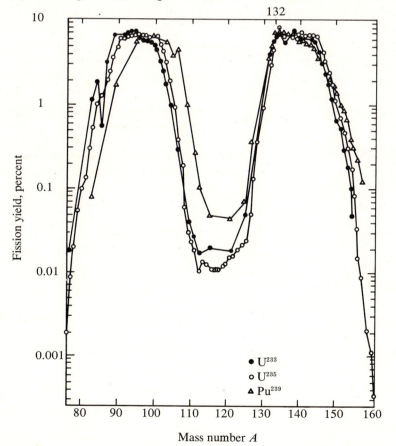

Table 5–4   Average energy release in fission of U^{235}†

Kinetic energy of fission fragments ($A \approx 95$ and 140)	165 ± 5 Mev	
Kinetic energy of prompt and delayed neutrons (2-3 neutrons)	5	
Prompt gamma rays (≈ 5 gamma rays)	6 ± 1	13
Beta rays (≈ 7 beta rays)	8 ± 1.5	7
Antineutrinos	12 ± 2.5	10
Radioactive gamma rays	6 ± 1	
Total energy release Q(overall)	204 ± 7 Mev	

† By permission from Segrè, 1964, chap. 11, sec. 11.

5-7b Details of the fission process.   The original theory of the fission process was developed by Bohr and Wheeler (1939) on the basis of the liquid-drop model. The process is now envisaged to occur as shown on Fig. 5-32. The binding energy of the captured neutron sets the compound nucleus into violent vibrations which break up the nucleus. Prompt neutrons are released. Some fission fragments are formed in excited states which decay by gamma radiation with

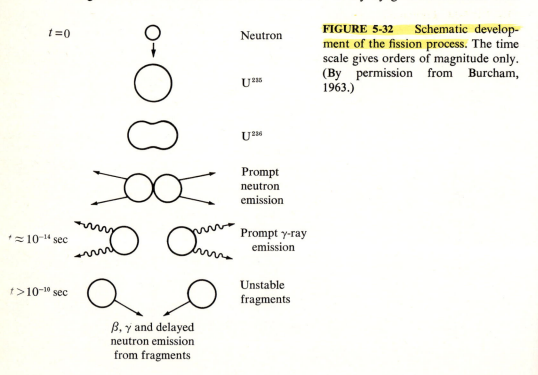

$t = 0$ Neutron

U^{235}

U^{236}

Prompt neutron emission

$t \approx 10^{-14}$ sec   Prompt γ-ray emission

$t > 10^{-10}$ sec   Unstable fragments

β, γ and delayed neutron emission from fragments

FIGURE 5-32   Schematic development of the fission process. The time scale gives orders of magnitude only. (By permission from Burcham, 1963.)

FIGURE 5-33 Calculated prompt energy release in symmetric fission [Eq. (5-91)].

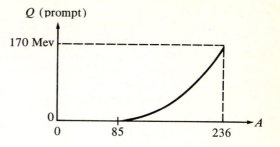

typical lifetimes of 10^{-15} to 10^{-13} sec, after which most fission fragments decay by negative beta decay towards the stability line.

If we calculate the energy released (5-91) when a nucleus (A,Z) is converted into two nuclei $(\tfrac{1}{2}A,\tfrac{1}{2}Z)$, which is called *symmetric fission,* we find[1] from the semi-empirical mass formula that Q(prompt) is positive for nuclei with $A > 85$ (Fig. 5-33). Yet such light nuclei do not fission spontaneously. We must therefore conclude that there is a *fission barrier.*

The best way to recognize this is to reverse the fission process. Assume that two spherical nuclei, each $(\tfrac{1}{2}A,\tfrac{1}{2}Z)$, are brought together as shown in Fig. 5-34a.

[1] Evans, 1955, p. 386.

FIGURE 5-34 Fission barrier. (*a*) Symmetric fission fragments. (*b*) Corresponding potential energy diagram. The approximate shape of the system is given below the abscissa. In case 1, the nucleus would promptly decay by spontaneous fission. In case 2, a certain amount of activation energy E_{ex} has to be furnished to produce fission. (After Evans, 1955.)

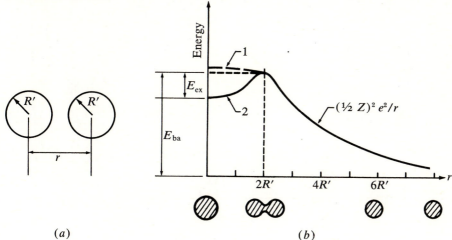

(*a*) (*b*)

FIGURE 5-35 Low-energy neutron fission cross section $\sigma(n,f)$ of U^{235}. For comparison, the total neutron cross section σ_{tot} of U^{235} is shown also. The difference between the two cross sections is mainly $\sigma(n,\gamma)$. (By permission from D. J. Hughes and J. A. Harvey, "Neutron Cross Sections," 1st ed. and J. R. Stehn et al., "Neutron Cross Sections," 2d ed., suppl. no. 2, vol. 3, Sigma Center, Brookhaven National Laboratory, Upton, New York, 1955 and 1965.)

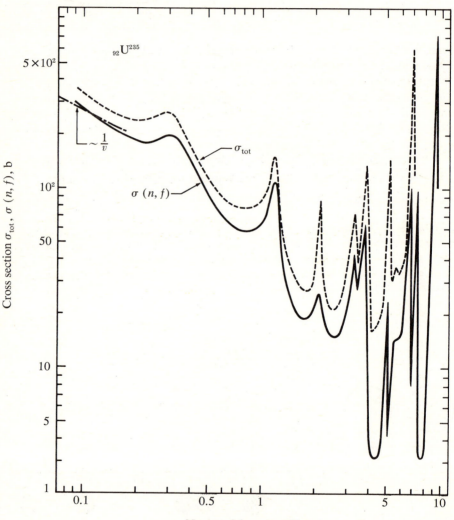

The potential energy between the spheres is equal to $(\frac{1}{2}Z)^2 e^2/r$, where r is the distance between the centers. When the spheres touch (Fig. 5-34b), nuclear forces begin to act and the spheres will coalesce. Two cases can occur:

1 The potential energy never decreases as the system takes on its shape of least distortion, i.e., a spherical shape.

2 The potential energy does decrease as the system takes on a spherical shape.

Conversely, starting with the spherical nucleus, case *1* leads to spontaneous fission; case *2* gives fission only if a certain excitation energy is provided,[1] which is denoted by E_{ex} in Fig. 5-34b. The energy release Q(prompt) is approximately equal to the barrier height E_{ba}. According to Bohr and Wheeler, the shape of the fissioning nucleus first becomes ellipsoidal. On this basis one can calculate the form of the potential curve in Fig. 5-34b near $r = 0$ and show[2] that case *1* is obtained only for $Z > 115$. Known nuclei, therefore, do not undergo spontaneous fission as their major mode of decay.

In a nucleus like U^{235}, the excitation energy E_{ex} (5-6 Mev) is furnished by the binding energy of the captured neutron (≈ 7 Mev). Fission therefore occurs with thermal neutrons. In the case of U^{238}, only about 5 Mev are gained when a thermal neutron is captured. This is less than E_{ex}, and the nucleus will fission only with fast neutrons. The difference in neutron binding energies is caused by the pairing term δ in Eq. (2-127). Hence, most fissionable even-even nuclei have *fission thresholds*, whereas most odd-A nuclei can fission with thermal neutrons.

5-7c Fission cross section. Because fission proceeds through a compound nucleus, the fission cross section is expected to follow Eq. (5-64) with $\Gamma_a = \Gamma_n$ and $\Gamma_b = \Gamma_f$, where Γ_f is called the *fission width*. This quantity is proportional to the probability that a given compound-nucleus level decays by fission.

The fission cross section of U^{235} is shown in Fig. 5-35. The thermal neutron part of this cross section follows the $1/v$ law [Eq. (5-74)]. At higher energies compound-nucleus resonances occur. The other major cross section which contributes to the total neutron cross section of U^{235} is neutron capture [$U^{235}(n,\gamma)U^{236}$]. The elastic scattering cross section at ev energies is approximately 10 b, which is very close to the value $4\pi R^2$ expected from Eq. (5-65). For the lowest resonance, shown in Fig. 5-35, the experimental parameters are

$$T_0^* = 0.29 \text{ ev} \qquad \Gamma_\gamma = 0.035 \text{ ev}$$
$$J = 3^- \text{ or } 4^- \qquad \Gamma_n \approx 3 \times 10^{-4} \text{ ev}$$
$$\Gamma_f = 0.10 \text{ ev}$$

[1] This statement is somewhat oversimplified because there can be tunneling through the barrier in case 2 (Sec. 2-2g), i.e., spontaneous fission can occur in this case also, but the probability is very small.

[2] Evans, 1955, pp. 387ff.

The threshold behavior of the U^{238} fission cross section (Fig. 5-36) is determined mainly by the fission penetrability factor, which is very similar to Eqs. (4-86) and (4-94). The interesting steps at higher energies are caused by secondary processes. The lowest threshold is, of course, caused by the reaction

$$U^{238} + n \rightarrow U^{239*} \rightarrow \text{fission} \qquad (5\text{-}93)$$

The next process sets in when neutrons have sufficiently high energies to provide the fission activation energy of U^{238} by inelastic scattering

$$U^{238}(n,n')U^{238*} \rightarrow \text{fission} \qquad (5\text{-}94)$$

FIGURE 5-36   Fast-neutron fission cross section of U^{238}. Thresholds of various processes are indicated. (Adapted from J. R. Stehn et al., "Neutron Cross Sections," 2d ed., suppl. no. 2, vol. 3, Sigma Center, Brookhaven National Laboratory, Upton, New York, 1965.)

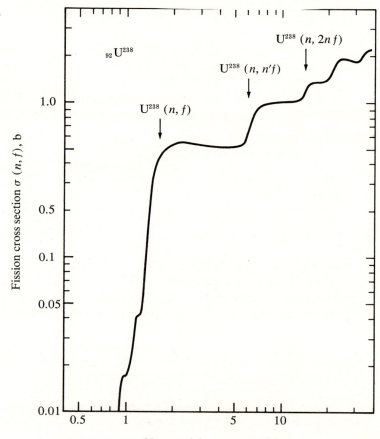

Another process can start when U^{237} can be created with sufficient excitation energy to fission

$$U^{238}(n,2n)U^{237}* \to \text{fission} \tag{5-95}$$

and so on.

A detailed examination of the lowest threshold has revealed finer steps in the U^{238} fission cross section (two are shown on Fig. 5-36). These steps have led to a more complete description of the fission process than given above.[1]

PROBLEMS

5-1 Using the notation of Sec. 5-2 prove, if Θ is the c.m. angle of emission of particle b, and θ the lab. angle, that

$$\cot \theta = \frac{(v_0/V_b) + \cos \Theta}{\sin \Theta}$$

and

$$\sin (\Theta - \theta) = \frac{v_0}{V_b} \sin \theta$$

5-2 Prove that Eqs. (5-14) and (5-15) are equivalent.

5-3 Show that in the reaction $X(a, b) Y$, the energy of particle b in the c.m. system is equal to

$$(M_Y/M)[Q + (1 - M_a/M)T_a]$$

where $M = M_b + M_Y \approx M_a + M_X$
$T_a = $ lab. kinetic energy of particle a
$Q = Q$ value of reaction.

5-4 (a) Compute the threshold energy for the $C^{12} + \gamma \to 3 He^4$ reaction. Use Appendix C. (b) If in this reaction, two of the alpha particles come off in the same direction with the same kinetic energy, what fraction of the available energy is carried off by the third alpha particle?

5-5 (a) Compute the Q values of the $H^2 + H^2 \to He^3 + n$ reaction and of the $H^3 + H^2 \to He^4 + n$ reaction. (b) Assume an electrostatic accelerator is available which will accelerate any particle of charge e to 4 Mev. What is the maximum neutron energy which could be produced using this accelerator in combination with either of the above reactions? (The mass of H^3 is 3.016050 u. The other masses can be looked up in Appendix C.)

5-6 (a) It is desired to obtain neutrons with a maximum energy of 2.0 Mev by bombarding tritium with protons. What must the energy of the protons be? [The threshold energy for the $H^3(p,n) He^3$ reaction is 1.019 Mev.] (b) For the conditions of part (a), what is the minimum energy of the neutrons emitted? (c) In which direction are the neutrons in parts (a) and (b) emitted, relative to the incident proton beam? [Answer this question first for yourself before starting (a) and (b).]

[1] Wheeler, 1963.

5-7   The $H^3(p,n)He^3$ reaction has a threshold energy of 1.019 Mev.   **(a)** If H^3 is bombarded with 1.100-Mev protons, what are the energies of the neutrons produced at $0°$ (in the forward direction)?   **(b)** If H^3 is bombarded with 1.019-Mev protons, what are the energy and direction of the neutrons produced?

5-8   The reaction $Li^7(p,n)Be^7$ $(Q = -1.64$ Mev$)$ is widely used for the production of monoenergetic neutrons.   **(a)** What is the maximum neutron energy which can be produced with a 3-Mev proton accelerator?   **(b)** If 3-Mev protons are used, what is the angle with respect to the proton beam axis, at which neutrons of 1.0-Mev energy are emitted?

5-9   The reaction of Prob. 5-8 has a differential cross section of 50 mb/sr at $0°$ and at a bombarding energy of 3.0 Mev. If a Li^7 target is 50 kev thick to 3.0 Mev protons, what is the number of neutrons/sec/sr emitted in the forward direction for a bombarding beam current of 1 μA? The energy loss of Li for 3-Mev protons is 100 kev-cm²/mg.

5-10   A rectangular metallic cell of dimensions $\frac{1}{2}$ cm $\times$ $\frac{1}{2}$ cm $\times$ 1 cm has a very thin foil window over one of the $\frac{1}{2}$-cm $\times$ $\frac{1}{2}$-cm square surfaces. The cell contains pure tritium at NTP. A 1-μA beam of 3.0-Mev protons enters the window parallel to the long side of the cell. Calculate the total yield of neutrons/sec. The cross section for the $H^3(p,n)He^3$ reaction is 0.50 barn for 3.0-Mev protons. (Neglect the energy loss of the proton beam in the foil and in the gas. Assume that tritium is a perfect gas.)

5-11 N^{14} has excited states at 2.31 and 3.95 Mev. If N^{14} gas is bombarded with 5.00-Mev neutrons, what are the energies of neutrons appearing at $90°$ with respect to the incident direction?

5-12   The $S^{32}(n,\alpha)Si^{29}$ reaction has a prominent resonance at 2.80-Mev neutron lab. energy (see Fig. 5-22a).   **(a)** Does this indicate a virtual state in the compound or in the final nucleus?   **(b)** What is the energy of this state above the ground state? (Use Appendix C.)

5-13   **(a)** Compute the classical distance of closest approach for a head-on collision in the scattering process appropriate to Fig. 5-28.   **(b)** Is this distance larger or smaller than the nuclear radius of Fe^{58}?   **(c)** Compute the Rutherford scattering cross section at the c.m. angles of $10°$ and $70°$.

5-14   The reaction $C^{13}(d,p)C^{14}$ $(Q = 5.95$ Mev$)$ has a resonance at a deuteron (lab.) energy of 2.45 Mev. Can you predict from this whether or not the reaction $B^{11}(\alpha,n)N^{14}$ $(Q = 0.15$ Mev$)$ may be expected to have a resonance and at which alpha (lab.) energy this would occur? C^{14} decays to N^{14} with $Q_\beta = 0.16$ Mev.

5-15   **(a)** The effect of a (d,p) reaction is to add a neutron to the target nucleus. Show that the binding energy of the last neutron in the product nucleus is given by the sum of the Q value of the (d,p) reaction and the binding energy of the deuteron.   **(b)** The reactions $Pb^{207}(d,p)Pb^{208}$ and $Pb^{208}(d,p)Pb^{209}$ have Q values of 5.14 and 1.64 Mev, respectively. What are the binding energies of the last neutron in Pb^{208} and Pb^{209}?   **(c)** Can you explain the difference between these binding energies on the basis of some nuclear model?

5-16   A state in C^{12} at 17.2-Mev excitation energy can decay by emission of either a proton or an alpha particle. The total width of the state is 1.16 Mev. The reaction

$B^{11}(p,\alpha)Be^8$ has a peak cross section of 0.16 b at a (lab.) proton energy of 1.4 Mev, which corresponds to the excitation of the 17.2-Mev state of C^{12}. Neglecting all spin factors, what can you say about the partial widths Γ_α and Γ_p from these data?

5-17 The $U^{235}(n,\gamma)U^{236}$ reaction has a resonance at an energy $T_0^* = 0.29$ ev (see Fig. 5-35). Appropriate data are given in Sec. 5-7c. **(a)** Compute the ratio $\sigma(n,n)/\sigma(n,\gamma)$ at the resonance. **(b)** Compute the magnitude of $\sigma(n,\gamma)$ at the resonance. **(c)** Compute the neutron reduced width of the resonance. **(d)** Compute the lifetime of this level.

5-18 Show that the shape of the $Co^{59}(p,n)$ cross section near threshold follows approximately Eq. (5-76). The cross section is shown in Fig. 5-23. The threshold energy of the reaction is 1.89 Mev.

5-19 **(a)** Compute the expression $\int_0^\infty \sigma(a,b)\, dT_0$ for Eq. (5-64), assuming all the widths to be constant. This is known as a resonance integral and has applications in nuclear reactor theory. **(b)** Compute the resonance integral for the (n,γ) resonance in U^{235} described in Sec. 5-7c. For simplicity take the widths as constants, even though this is not a good assumption here.

5-20 Analyze Fig. 5-28 on the basis of Eq. (5-89). **(a)** Is the relation between Θ and n satisfied? **(b)** What value of λ do you obtain, and does it make sense? [Because Eq. (5-89) is completely oversimplified, you should not expect more than qualitative agreement.]

5-21 If the two fission fragments of $(U^{235} + n)$ fission have the mass numbers and total kinetic energy given in Table 5-4, what are their individual kinetic energies? Is your answer accurate, and if not, why not?

5-22 Compute from the semiempirical mass formula the value of A at which Q(prompt) [Eq. (5-91)] is zero for symmetric fission. (See Fig. 5-33.)

5-23 Suppose U^{235} fissions into two fragments with $A = 91$ and $A = 139$, plus several neutrons. **(a)** What is the potential energy between the two fission fragments at the point at which they have just separated? **(b)** If these fragments decay only by beta and gamma decay, what would you expect the stable end products of the fission chains to be?

5-24 A beam of 0.1-ev neutrons bombards a 1-cm cube of natural uranium metal. If the flux of the beam is 10^{12} neutrons/sec/cm^2, what is the rate generation of heat in the block due to the slow-neutron fission of U^{235} (natural abundance 0.72 percent)? Use information from Fig. 5-35 and Table 5-4. Conversion factors are given in Appendix D.

NUCLEAR FORCE 〜〜〜〜〜〜〜〜〜〜〜〜〜〜 6

6-1 INTRODUCTION

As we mentioned at the beginning of the book, the two central problems in nuclear physics are, first, to understand the nature of the force acting between nucleons and, second, to explain the properties of a complex nucleus (many-nucleon system) in terms of the nuclear force. These problems, although obviously related, are essentially different, for even if the nuclear force were known perfectly, there would still be the problem of dealing with a many-particle system. This is present even in classical physics.

In Chaps. 2 and 5 we saw how one can understand properties of a complex nucleus in terms of nuclear models. These models require that, at least in a complex nucleus, the nuclear force has the following properties:

1 There is a dominant short-range part, which is central and which provides the overall shell-model potential.

2 There is a part whose range is much smaller than the nuclear radius, which tends to make the nucleus spherical and to pair up nucleons.

3 There is a part whose range is of the order of the nuclear radius, which tends to distort the nucleus.

4 There is a spin-orbit interaction.

5 There is a spin-spin interaction.

6 The force is charge independent.

7 The force saturates.

More information concerning the force between two nucleons can be obtained from the simplest two-nucleon system, the deuteron, and from proton-proton as well as neutron-proton scattering. On the other hand, neutron-neutron scattering cannot be investigated with presently available neutron fluxes.[1] The interpretation of experiments, such as those discussed in Appendix A, supports the short-range aspects (property 1) of the dominant part of the nuclear force. The range is about 2 F and an attractive potential of approximately 30 Mev is found, if the force is represented by a potential interaction. The properties 4, 5, and 6, given above, are also confirmed, although it appears that charge independence of the nuclear force is not perfect.

Two other properties, which are helpful in explaining saturation (property 7, see Sec. 2-3b), can be deduced from high-energy nucleon-nucleon scattering. At a distance of about $\frac{1}{2}$ F, the nucleon-nucleon force becomes very repulsive; pictorially, we speak of a hard core. Also, there is a force which can change a neutron into a proton during a collision. This is called *exchange force* and will be discussed further below. It is many orders of magnitude stronger than the beta-decay interaction, which can also change a neutron into a proton, and vice versa.

6-2 MESON THEORY OF NUCLEAR FORCES

In modern physical theories, any attractive force between two particles is regarded as the exchange of an attractive property. The following example can serve as an illustration. Consider two protons, separated by a distance of the order of 10^{-8} cm. They will repel each other by the coulomb force. If an electron is placed in their neighborhood (Fig. 6-1a), both protons will be attracted to the

---

[1] In this connection, we remark that the *free* two-nucleon force may not be identical in all respects to the two-nucleon force inside a complex nucleus. So far, there is no evidence available that the forces are different.

FIGURE 6-1   Illustration of an exchange force between two protons: (a) in an $(H_2)^+$ molecule,   (b) in a nucleus.

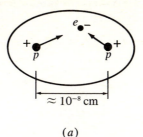

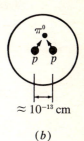

$\approx 10^{-8}$ cm

$\approx 10^{-13}$ cm

(a)   (b)

electron. In fact, the force of attraction is sufficiently strong to overcome the original repulsion, and a stable $(H_2)^+$ molecule can be formed. In this example, the exchanged attractive property between the two particles is an electron.

If a third proton is placed in the neighborhood of an $(H_2)^+$ molecule, the system will not be stable. This is caused by the Pauli exclusion principle. In the lowest-energy state, which is a $1s$ state, the two original protons must have antiparallel intrinsic spins. A third proton, placed into the same state, would violate the Pauli exclusion principle. If it is placed into a higher energy state, the average distance of separation increases (compare with Fig. 2-24), and the exchange force is very much weakened. Therefore an exchange force can saturate. Indeed, the original explanation of the saturation of nuclear forces (Sec. 2-3b) was made in terms of exchange forces (Heisenberg, 1932), although currently the hard core in the nuclear force is also recognized as an important contributor to saturation.

Yukawa (1935) first suggested that some heavy particle, later found to be a π meson, also called *pion*, should be exchanged between two nucleons in a nucleus (Fig. 6-1b) in order to provide binding with a sufficiently short range. Although at first he assumed that only charged mesons are exchanged between nucleons, later neutral mesons were included in the theory, and indeed π^+, π^0, and π^- mesons are now known to exist.

To show that the range of the force is related to the mass of the exchanged particle, consider the following model of the exchange mechanism between two protons (Fig. 6-1b). Assume that normally the π^0 meson is contained *virtually* in one of the protons. The mass of this object is M_p, the mass of a proton. Now suppose that from time to time the proton dissociates into a real π^0 meson and a proton. The mass of such an object would be $M_p + m_\pi$ where m_π is the mass of a π^0 meson (Fig. 6-2a). According to the uncertainty principle expression (4-32), a temporary dissociation would be allowable if it does not take a time longer than t, where

$$t \approx \hbar/\Delta E \qquad (6-1)$$

and from Fig. 6-2b

$$\Delta E = (M_p + m_\pi)c^2 - M_p c^2$$

$$= m_\pi c^2 \qquad (6-2)$$

FIGURE 6-2 Dissociation of a proton into a proton and π° meson. (a) Pictorial diagram. (b) Energy diagram.

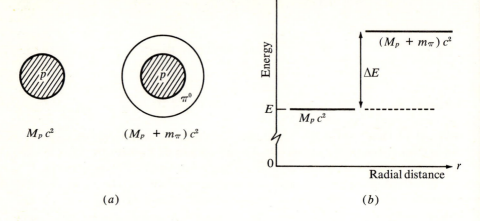

(a)

(b)

The greatest distance the meson could travel in this time is

$$r(\text{max}) \approx ct \tag{6-3}$$

where c is the speed of light. Hence

$$r(\text{max}) \approx \frac{\hbar}{m_\pi c} \tag{6-4}$$

This gives an estimate for the range of the pion exchange force.

The same expression can be derived by assuming that the wave function of the π° meson, in the region where it is far separated from the proton, is given by the Schrödinger equation (2-47) with $l = 0$

$$-\frac{\hbar^2}{2m_\pi}\frac{d^2u}{dr^2} = (E - V)u$$

$$= -\Delta E u \tag{6-5}$$

The corresponding solution is of the form

$$u \to a e^{\kappa r} + b e^{-\kappa r} \tag{6-6}$$

where

$$\kappa = \frac{(2m_\pi \Delta E)^{\frac{1}{2}}}{\hbar} \tag{6-7}$$

in complete analogy to the one-dimensional equations (2-99) and (2-100). Since the system is bound, $u(r \to \infty) = 0$ and $a = 0$ [see Eq. (2-42)]. At large distances from the proton the π^0 wave function is therefore [Eq. (2-48)]

$$\psi \approx \frac{b}{r} e^{-\kappa r} \qquad (6\text{-}8)$$

where

$$\kappa = \frac{m_\pi c}{\hbar} \qquad (6\text{-}9)$$

A factor $\sqrt{2}$ has been omitted in Eq. (6-9) in accordance with the result for ψ obtained by using a relativistic wave equation,[1] rather than Eq. (6-5). Another proton, brought into the neighborhood of the first one, will feel the effect of the π^0-meson "cloud" (6-8) and under reasonable assumptions[1] the strength of the interaction between the two protons is proportional to expression (6-8). This is called the *Yukawa potential.* Its "range" is given by $1/\kappa$, which is identical to Eq. (6-4). Substituting the known mass of the π meson ($\approx 270 m_e$), we find

$$\frac{1}{\kappa} \approx 1.4 \text{ F} \qquad (6\text{-}10)$$

which is of the right magnitude to account for the range of the nuclear force.

Detailed analysis of the high-energy nucleon-nucleon scattering has shown that at large distances ($r > 2$ F) the radial dependence of the nuclear interaction is correctly given by expression (6-8). We can account for charge independence by assuming π^+, π^0, and π^- mesons are exchanged between nucleons, i.e., that within each nucleon the following dissociation takes place from time to time (compare Fig. 6-2a)

$$\begin{array}{ll} p \rightarrow p + \pi^0 & p \rightarrow n + \pi^+ \\ n \rightarrow n + \pi^0 & n \rightarrow p + \pi^- \end{array} \qquad (6\text{-}11)$$

This dissociation also explains the charge-exchange process in a high-energy nuclear collision between a neutron and a proton: a charged π meson is transferred from one to the other.

It has been possible to investigate the charge cloud around nucleons by high-energy electron scattering (Hofstadter et al., 1960; Littauer et al., 1961). The picture implied by the processes (6-11) appears to be correct: the outer charge of a proton is positive, and that of a neutron predominantly negative (Fig. 6-3).

For reasons that would take us too far afield, the spin-orbit part of the nuclear force cannot be caused by[1] the processes (6-11). Also, the shorter-range ($r < 2$ F) properties of the nuclear force must be influenced strongly by the exchange of other mesons or by multiple pion exchange. The same reasoning which leads to Eq. (6-4) shows that if n pions are exchanged between nucleons, the range decreases by roughly $1/n$. Particles which decay predominantly into

[1] Brink, 1965, chap. 6.

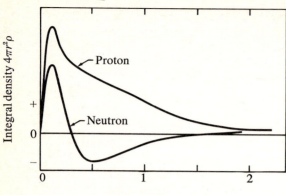

FIGURE 6-3 Radial charge distribution in the proton and neutron versus radial distance from the center. The charge density is denoted by ρ and $4\pi r^2 \rho$ is the total charge between r and $r + dr$. (Adapted from Littauer, Schopper, and Wilson, 1961.)

two or more pions have, in fact, been found and may play a role in determining the nuclear force at distances of separation less than 2 F.

If indeed the proton and neutron have the complicated structure implied by the processes (6-11), we might expect to find excited states of the pion-nucleon system. Figure 6-4 shows that pion-proton total cross sections indeed have resonances. From the discussion in Chap. 5 it should be clear that these resonances reflect excited states of the compound system. The nucleon therefore

FIGURE 6-4 Total proton-pion cross section versus total energy of the compound system. (By permission from Segrè, 1964.)

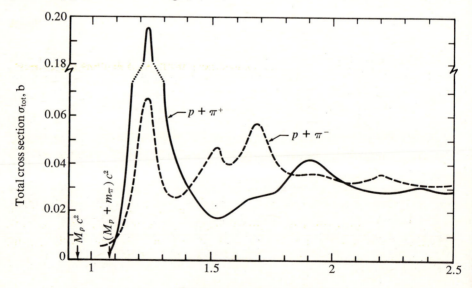

FIGURE 6-5 Excited states of a nucleon, inferred from Fig. 6.4 and other experiments. [By permission from A. H. Rosenfeld et al., *Rev. Mod. Phys.* **37**: 633 (1965).]

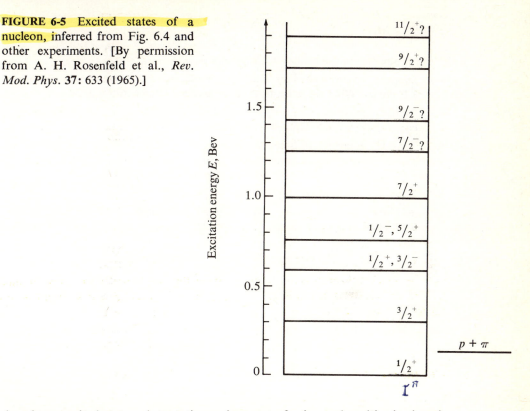

does have excited states. A tentative assignment of spins and parities is given in Fig. 6-5.

Pursuit of this interesting topic is beyond the scope of the present treatment. It does, however, cause us to reflect whether all physical systems are complex. As time progresses and experimental tools become more refined, the search for more fundamental systems might actually lead to systems of progressively greater complexity. First, man was faced with the complexity of the solar system, into which order was brought by classifying the motion of the planets. Next, the Mendeléef table provided an order for chemical complexity, which was clarified further by the electronic structure of atoms. At the nuclear level, the proton-neutron structure indicates another hierarchy of complexity. Now, the nucleons themselves are found to be complex structures and nearly 100 other unstable particles are known. We may well speculate whether the hierarchies of complexity will continue to increase. Or will we finally find particles which create each other by their mutual interactions and which will form the basic substance from which all other particles are made?[1]

---

[1] For a lucid description of the present state of knowledge and speculation see Foldy, 1965.

PROBLEMS

These problems require a study of Appendix A.

6-1 Suppose the proton-pion interaction is represented by a square well potential of 1 F radius. What is the depth of the potential? (Refer to Fig. 6-2.)

6-2 Compute an exact value for the rms radius of the deuteron, Eq. (A-13), using the correct *internal* and *external* wave functions.

6-3 Ten-Mev neutrons are scattered by He^4. What is the highest partial wave, i.e., that of largest angular momentum, which you expect to be influenced by the nuclear interaction between the neutron and the He^4 nucleus?

6-4 Prove Eq. (A-31), starting with Eq. (A-29).

6-5 (a) Show that for a square well of depth V_0 and range r_0, the scattering length a for a *spinless* neutron is given by the relation (M_0 = reduced mass of neutron)

$$K_0 \cot K_0 r_0 = \frac{1}{r_0 - a} \qquad \text{where } K_0 = (2M_0 V_0)^{\frac{1}{2}}/\hbar$$

(b) Compute a for $V_0 = 36$ Mev, $r_0 = 2.0$ F, and *n-p* scattering.

6-6 Prove Eq. (A-57).

NUCLEAR FORCE INFORMATION FROM ▨▨▨▨▨▨ A THE TWO-NUCLEON SYSTEM

The methods of extracting information about the nuclear force from the properties of the two-nucleon system are fairly complicated[1]. Nevertheless, an understanding of the methods can be given insofar as they require only a sophistication in quantum mechanics consistent with the rest of our presentation.

The two-nucleon system allows us to apply to the simplest *complex* nucleus H^2 the ideas presented in Chaps. 2 and 5. We can study on the one hand the level structure of H^2 and on the other hand a nuclear reaction (neutron-proton scattering) involving the same compound system. The problem is calculable in detail, because with only two nucleons interacting, the potential

[1] Brink, 1965.

between the nucleons, which gives the level structure, also describes the nuclear reaction in a simple fashion (Wigner, 1933). By comparing the n-p system with the p-p system and the n-n system (possible only indirectly), one can also obtain information about the charge independence of the nuclear force (Sec. 2-7).

A-1 STRUCTURE OF THE DEUTERON

The binding energy of the deuteron is 2.23 Mev. It has been determined by measuring the energy of the gamma rays emitted in the thermal neutron capture by protons

$$n + \mathrm{H}^1 = \mathrm{H}^2 + \gamma$$

The inverse reaction has also been used, employing electrons of known energy to produce external bremsstrahlung for the photodisintegration of the deuteron. No stable excited states of the deuteron have been found. Various reactions, including n-p scattering, have led to the discovery of a virtual state (see Fig. 2-29) at approximately 70-kev energy above the n-p breakup threshold, i.e., at an excitation energy of 2.30 Mev above the ground state.

To gain some insight about extracting nuclear force information from the deuteron level structure, we make the simplest possible assumption about the force. We assume that the force is central and caused by a potential of the form

$$V(r) = \begin{cases} -V_0 & \text{for } r < r_0 \\ 0 & \text{for } r > r_0 \end{cases} \tag{A-1}$$

where r is the internucleon distance. This is called a *square well potential*. It has a constant value throughout a spherical volume of radius r_0. Even though this shape of potential is very much oversimplified in comparison with the true nature of the nuclear force (Sec. 6-2), we calculate the level structure of a system described by this potential. We use the radial Schrödinger equation (2-47) with the substitution

$$m_0 = \tfrac{1}{2}M \tag{A-2}$$

for the reduced mass [Eq. (2-62)]. Consistent with the approximations we will make, the proton mass is set equal to the neutron mass and both are denoted by M.

There is good evidence that the ground state of the deuteron is a $1s$ state, i.e., has $l = 0$. First, the lowest-energy state in practically any potential is an s state[1] (see Figs. 2-23 or 2-25 and also recall the ground state of the hydrogen atom). Second, the magnetic moment of the deuteron is nearly equal to the algebraic sum of the proton and neutron moments, indicating that the intrinsic

---

[1] This is because the $1s$-wave function has the smallest curvature and, hence, the smallest value of $-d^2u/dr^2$, which is related to the average kinetic energy in quantum mechanics [compare Eqs. (2-47) and (2-51)]. The average potential energy is less dependent on the shape of the wave function and therefore the kinetic energy often determines the total energy.

spins of these particles are aligned and that there is no orbital motion of the proton with respect to the neutron.[1] This is consistent with the total angular momentum $I = 1$ of the deuteron ground state.

For $l = 0$, the s-wave Schrödinger equation (2-53) can be used for the radial wave function $R(r) = u(r)/r$. Substituting Eq. (A-2)

$$-\frac{\hbar^2}{M}\frac{d^2u}{dr^2} + V(r)u = Eu \tag{A-3}$$

As we mentioned in Sec. 2-2d, this equation is mathematically equivalent to the one-dimensional equation (2-22), except for the additional requirement $u(0) = 0$ [Eq. (2-54)]. We divide the radial space into the regions $r \leq r_0$ where $V = -V_0$ and $r > r_0$ where $V = 0$. For the ground state of the deuteron, we can set $E = -B$, where B is the binding energy of the deuteron. We find for $r \leq r_0$

$$u = ae^{iKr} + be^{-iKr} \quad \text{where} \quad K = [M(V_0 - B)]^{\frac{1}{2}}/\hbar \tag{A-4}$$

and for $r > r_0$

$$u = a'e^{\kappa r} + b'e^{-\kappa r} \quad \text{where} \quad \kappa = (MB)^{\frac{1}{2}}/\hbar \tag{A-5}$$

Reference to Sec. 2-2g shows the mathematical correctness of these general solutions. Physically, the following boundary conditions must be imposed:

1 $u(0) = 0$, to keep $R(0)$ finite.
2 $u(r \to \infty) = 0$, since we are dealing with a bound state [Eq. (2-42)].
3 At $r = r_0$, the values and derivatives of the functions (A-4) and (A-5) must match.

Condition *1* gives $a = -b$, so that we can write for $r \leq r_0$

$$u = c \sin Kr \tag{A-6}$$

where c is a new constant. Condition *2* gives $a' = 0$, so that for $r > r_0$

$$u = b'e^{-\kappa r} \tag{A-7}$$

Condition *3* gives,[2] after eliminating c and b',

$$K \cot Kr_0 = -\kappa \tag{A-8}$$

or, with the help of Eqs. (A-4) and (A-5)

$$\tan Kr_0 = -\frac{K}{\kappa} = -\left(\frac{V_0 - B}{B}\right)^{\frac{1}{2}} \tag{A-9}$$

---

[1] Orbital motion of the proton would have a magnetic effect similar to that of a current loop and so produce an additional magnetic moment.
[2] The same relation can be obtained by matching the quantity $(1/u)(du/dr)$, called *logarithmic derivative*, at $r = r_0$ instead of matching u and du/dr separately. The logarithmic derivative is a useful quantity because it also occurs in reaction theory.

We will find later that $V_0 \approx 36$ Mev, so that[1] with $B = 2.23$ Mev,

$$Kr_0 \approx \tfrac{1}{2}\pi \qquad\qquad (\text{A-10})$$

Using Eq. (A-4) once more, we obtain

$$V_0 r_0{}^2 \approx \left(\frac{\pi}{2}\right)^2 \frac{\hbar^2}{M} \approx 1.0 \text{ Mev-barn} \qquad\qquad (\text{A-11})$$

In other words, from the deuteron binding energy alone, we cannot determine r_0 and V_0, but only the above combination. An examination of the n-p scattering cross section (Sec. A-3) gives $r_0 \approx 2F$, yielding $V_0 \approx 36$ Mev from Eq. (A-9).

FIGURE A-1 (a) Square well potential for the bound state of the deuteron. (b) Corresponding radial wave function $u(r) = rR(r)$.

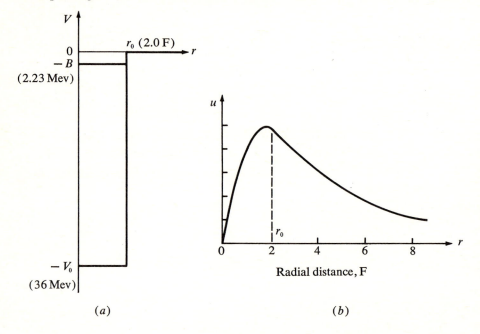

(a)   (b)

The wave function $u(r)$ is shown in Fig. A-1. Inside the nuclear potential, the wave function is approximately one-quarter of a sine wave. Because it must match a decreasing exponential for $r > r_0$, Kr_0 must be slightly larger than 90° [see footnote preceding Eq. (A-10)]. The relatively slow falloff of the wave function, characterized by the $1/e$-falloff distance [Eq. (A-7)]

$$\frac{1}{\kappa} = \frac{\hbar}{(MB)^{\frac{1}{2}}} = 4.3 \text{ F} \qquad\qquad (\text{A-12})$$

[1] A more accurate value for Kr_0 from Eq. (A-9) is 104° rather than 90°.

means that the two nucleons in the deuteron spend a large fraction of their time at separations $r > r_0$. This is the classically forbidden region of *negative kinetic energy*, which we encountered in Sec. 2-2g. If we define a mean-square nuclear radius R_{rms} for the deuteron by

$$R_{rms}^2 = \frac{\int_0^\infty r^2 R^2(r) r^2 \, dr}{\int_0^\infty R^2(r) r^2 \, dr} \tag{A-13}$$

substitution of the wave function (A-7) for the entire region produces a rough estimate and gives

$$R_{rms} = \frac{\hbar}{(2MB)^{\frac{1}{2}}} = 3.0 \text{ F} \tag{A-14}$$

This can be compared with the value 2.0 F found by electron scattering (Sec. 1-2b).

For the virtual state of the deuteron, the total angular momentum is zero, and we turn to the *n-p* scattering problem to find out more about the state. Since, relatively speaking, the virtual state lies close to the ground state, we also expect its wave function inside the appropriate nuclear potential to be approximately equal to a one-quarter sine wave.[1] For the moment, we leave it an open question whether the range or strength, or both, of the potential needed to describe the virtual state are different from the ground-state potential.

A-2 SCATTERING THEORY

Scattering amplitude.   To see how the potential influences *n-p* scattering, we need to consider some parts of quantum-mechanical scattering theory.[2] In essence we are dealing with a three-dimensional problem very similar to the one-dimensional problem of Sec. 2-2g, where we considered the reflection (i.e., *scattering*) and transmission of a beam of particles by a potential barrier.

We consider a single scattering nucleus, represented as in Fig. A-2 by a potential V. For simplicity, we assume that V is spherically symmetric so that it depends only on the radial distance r of the scattered particle from the center of the scatterer. Our aim will be to compute the differential scattering cross section $d\sigma/d\Omega$ using Eq. (5-35). For this purpose we must find the number of particles dN_1 scattered in unit time by one target nucleus into a solid angle $d\Omega$ (Fig. 5-9), and we must compute the incident flux F_{in} [Eq. (5-26)]. Then,

$$\frac{d\sigma}{d\Omega} = \frac{dN_1/d\Omega}{F_{in}} \tag{A-15}$$

---

[1] The virtual state cannot be the 2s state in the potential $- V_0$. For such a state, the wave function would be approximately a three-quarter sine wave within $r < r_0$. This triples the value of K [compare Eq. (A-10)] and puts the excitation energy at nine times the value $(V_0 - B)$, i.e., at about 300 Mev above the bottom of the potential.

[2] Schiff, 1955, sec. 18.

FIGURE A-2 Quantum-mechanical scattering problem. The total wave function far from the scattering potential $V(r)$ consists of an incident plane wave Ψ_{in} and a scattered spherical wave Ψ_{sc}. The number of particles scattered per unit time into a small solid angle $d\Omega$ at a scattering angle θ is denoted by dN_1.

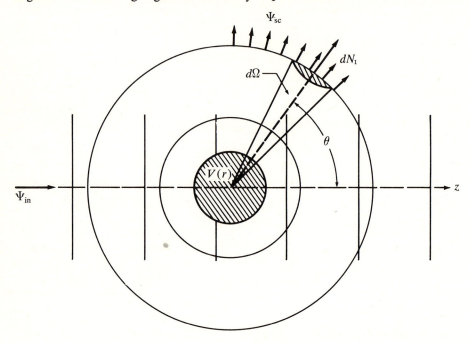

Far from the scatterer, the wave function consists of an incident part Ψ_{in}, representing the incoming beam of particles, and a scattered part Ψ_{sc}, representing the scattered particles. As in Eqs. (2-32) or (2-96)

$$\Psi_{in} = ae^{i(kz-\omega t)} \tag{A-16}$$

where $k = (2m_0 T_0)^{\frac{1}{2}}/\hbar$ = wave number of particles incident along z direction
$\quad m_0$ = reduced mass of incoming particle
$\quad T_0$ = c.m. kinetic energy of incoming particle

The angular frequency ω is given by Eq. (2-29), but will not be used in the problem because we look at the scattering process in a steady-state fashion. We assume there is a continuous supply of incoming particles (or, better, waves) and a continuous scattering away from the scattering center. The particles not affected by the scatterer go "to infinity" along the beam direction. According to Eq. (5-26) the incoming flux of particles is

$$F_{in} = \Psi_{in}^{*}\Psi_{in}v$$
$$= |a|^2 v \tag{A-17}$$

where v is the speed of an incoming particle with respect to the scatterer.

Far from the scatterer, the scattered particles will travel in a radial direction. They are therefore represented by a radially outgoing, traveling wave $e^{i(kr-\omega t)}$. But the total number of particles traversing a spherical surface surrounding the scatterer must be independent of the radius r of the surface. Hence, we write

$$\Psi'_{sc} = af(\theta)\frac{e^{i(kr-\omega t)}}{r} \qquad (A\text{-}18)$$

where a is included for convenience and $f(\theta)$ is an amplitude factor, independent of r, which must be calculated from the complete Schrödinger equation. The amplitude $f(\theta)$ is called the *scattering amplitude*. The number of particles dN_1 scattered per unit time into the solid angle $d\Omega$ (see Fig. A-2) is equal to the flux of scattered particles $\Psi'^{*}_{sc}\Psi'_{sc}v$ multiplied by the area $r^2\,d\Omega$ cut out by $d\Omega$ on a spherical surface of radius r.

$$dN_1 = \Psi'^{*}_{sc}\,\Psi'_{sc}\,vr^2\,d\Omega$$
$$= |a|^2\,|f(\theta)|^2\,v\,d\Omega \qquad (A\text{-}19)$$

Note that since we wrote Ψ'_{sc} in the form (A-18), dN_1 is independent of r. In a radial motion, the number of particles traveling within the cone defined by $d\Omega$ must be independent of r.

Substituting Eqs. (A-17) and (A-19) into Eq. (A-15) we find for the differential scattering cross section

$$\frac{d\sigma}{d\Omega} = |f(\theta)|^2 \qquad (A\text{-}20)$$

and for the integrated scattering cross section [Eq. (5-36)]

$$\sigma = \int |f(\theta)|^2\,d\Omega \qquad (A\text{-}21)$$

Our problem has now been simplified in the following way. If, far from the scatterer, we can find a solution of the (time-independent) Schrödinger equation (2-19)

$$-\frac{\hbar^2}{2m_0}\nabla^2\psi + V\psi = E\psi \qquad (A\text{-}22)$$

and can put it into the form

$$\psi = a[e^{ikz} + f(\theta)\,r^{-1}\,e^{ikr}] \qquad (A\text{-}23)$$

then the factor $f(\theta)$ can immediately be recognized as the scattering amplitude and used to compute the differential scattering cross section [Eq. (A-20)].

Partial wave analysis.[1]   Our task is made easier if we can write the solution (A-23) completely in the spherical coordinates r and θ, which we introduced in Sec. 2-2d. (For a spherically symmetric potential there can be no φ-angle dependence of the solution.) For this purpose, we decompose both

---

[1] Schiff, 1955, sec. 19.

terms in the right brackets of Eq. (A-23) into *partial waves*, each corresponding to an orbital angular momentum l (in units of $\hbar$).

From a semiclassical point of view, we are characterizing the particles in the incident beam by their impact parameters y, shown in Fig. 5-21, which can extend from zero to "infinity." Each impact parameter is related to l by Eq. (5-50)

$$y \approx l \lambdabar \tag{A-24}$$

where $\lambdabar = 1/k$ is the reduced de Broglie wavelength of the incident particles. Classically, y can be distributed in a continuous manner. In reality, though, l is quantized in integral steps so that the incident beam must be decomposed into zones, shown schematically in Fig. A-3. Each zone is characterized by a given l. We note in passing that this picture can be used in nuclear reaction theory to make an estimate of the total reaction cross section, i.e., the cross section due to all processes except shape elastic scattering. The area of each of the zones in Fig. A-3 is approximately

$$\pi(y_{l+1}^2 - y_l{}^2) = \pi \lambdabar^2[(l+1)^2 - l^2)] = (2l+1)\pi \lambdabar^2 \tag{A-25}$$

If all the particles up to a certain $l_{\max}$ react with the target nucleus and the other particles do not react at all

$$\sigma_{\text{react}} \approx \pi \lambdabar^2 \sum_0^{l\max} (2l+1) = \pi \lambdabar^2[(l_{\max}+1)^2)] \tag{A-26}$$

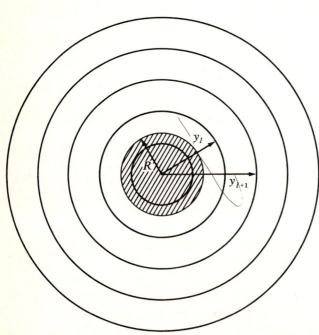

FIGURE A-3 Semiclassical view of partial wave analysis. The incident beam, seen in cross section on the figure, consists of particles of varying impact parameters y_l, related to the angular momentum l by Eq. (A-24). A nuclear interaction of radial extension R affects only those particles whose impact parameters lie approximately within R. This figure should be compared with Fig. 5-16, which shows a side view of the motion of a particle in the beam.

If l_{max} is given by Eq. (5-51) we find

$$\sigma_{react} \approx \pi(R + \lambda)^2 \qquad (A\text{-}27)$$

This can be used only for very crude estimates of the energy dependence of this cross section.

When we discussed the separated solution (2-43) of the Schrödinger equation in spherical coordinates, we mentioned that it was of the form

$$R(r)\, P_l^{(m)}(\cos\theta)\, e^{im\varphi} \qquad (A\text{-}28)$$

where l was the orbital angular momentum of the wave. Therefore, it seems reasonable that if we have a general solution of the Schrödinger equation $F(r, \theta)$ (which does not have a φ-angle dependence) and make a series expansion of the form[1]

$$F(r,\theta) = \sum_{l=0}^{\infty} F_l(r)P_l(\cos\theta) \qquad (A\text{-}29)$$

each coefficient $F_l(r)$ is associated with a definite orbital angular momentum l. For the incoming wave function $e^{ikz}(= e^{ikr\cos\theta})$, one can show that far from the origin $(r \gg 1/k)$

$$F_l(r) = i^l(2l + 1)\frac{\sin(kr - \tfrac{1}{2}l\pi)}{kr} \qquad (A\text{-}30)$$

and in particular for the s wave component of e^{ikz}

$$F_0(r) = \frac{\sin kr}{kr} \qquad (A\text{-}31)$$

For the scattering amplitude $f(\theta)$, we define (constant) partial wave scattering amplitudes f_l by the relation

$$f(\theta) = \sum_{l=0}^{\infty} f_l P_l(\cos\theta) \qquad (A\text{-}32)$$

The further simplification we have now accomplished is the following. Assume that we have solved the Schrödinger equation (A-22) *for a given l* and have the solution in the form (A-28) (with $m = 0$ to eliminate the φ angle, since we are considering only spherically symmetric potentials)

$$R(r)P_l(\cos\theta)$$

---

[1] $P_l(\cos\theta)$ is called a Legendre polynomial and is identical with the associated Legendre polynomial $P_l^{(0)}(\cos\theta)$. The first few polynomials are $P_0 = 1$, $P_1 = \cos\theta$, $P_2 = \tfrac{1}{2}(3\cos^2\theta - 1)$. They obey an *orthogonality relation*

$$\int P_l P_{l'}\, d(\cos\theta) = 0 \quad \text{if} \quad l \neq l' \quad \text{and} \quad \int P_l^2\, d(\cos\theta) = 2/(2l + 1)$$

This can be used to evaluate the coefficients $F_l(r)$ in particular cases by multiplying both sides of Eq. (A-29) by $P_{l'}$ and integrating over $d(\cos\theta)$.

far from the scatterer. This expression can now be compared with the lth component of expression (A-23). In particular, if we restrict ourselves to s waves, we find with the help of Eqs. (A-31) and (A-32)

$$R(r) = a\left(\frac{\sin kr}{kr} + f_0 \frac{e^{ikr}}{r}\right)$$

$$= \frac{a}{2ikr}[(1 + 2ikf_0)e^{ikr} - e^{-ikr}]$$

or $\qquad\qquad u(r) \sim (1 + 2ikf_0)e^{ikr} - e^{-ikr}$ \hfill (A-33)

where we concentrate just on the radial dependence of the wave function. By putting the radial wave function in this form, we can recognize the scattering amplitude f_0 by inspection. From Eq. (A-20) we can then compute the s-wave differential scattering cross section

$$\frac{d\sigma}{d\Omega} = |f_0|^2$$ \hfill (A-34)

which is independent[1] of the angle θ. Also, from Eq. (A-21)

$$\sigma = 4\pi|f_0|^2$$ \hfill (A-35)

We will show below that, far from the scatterer, the s-wave wave function[2] can always be put into the form (A-33).

S-wave phase shift.   We reasoned in Sec. 5-4d that only those bombarding particles with a given l (translate now: partial waves with a given l) should be affected by the nuclear interaction, for which $l \leq R/\lambda$. In the case of n-p scattering, R would have to be identified with the range r_0 (≈ 2 F) of the interaction, rather than with the radius of the deuteron. Looking at Table 5-2, we see that for c.m. energies below about 5 Mev (lab. energies below 10 Mev) the n-p interaction should take place only with $l = 0$. We therefore restrict our discussion henceforth to s waves.

In this case, the Schrödinger equation which has to be solved is again Eq. (A-3); but in the scattering problem, the energy E is positive and equal to the c.m. kinetic energy T_0 of the incoming particle. As we showed in the preceding section, we must find the solution far from the scatterer and put it into the form (A-33) to compute $d\sigma/d\Omega$. "Far from the scatterer" means that $V(r)$ should be zero there, which for the square well (A-1) applies for $r > r_0$. Hence, we need to solve only the equation

$$-\frac{\hbar^2}{M}\frac{d^2u}{dr^2} = T_0 u$$ \hfill (A-36)

---

[1] We used this fact in Sec. 3-3b.
[2] This can also be extended to other l values.

The most general solution is

$$u = A \sin (kr + \delta_0) \qquad \text{(A-37)}$$

where $k = (MT_0)^{\frac{1}{2}}/\hbar$, and A and δ_0 are constants which must be determined by the solution of Eq. (A-3) in the region where $V(r)$ is finite. The quantity δ_0 is called the *s-wave phase shift*. It is often used instead of f_0 to parameterize the differential cross section (A-34). Let us attempt to write Eq. (A-37) in the form (A-33)

$$u = \frac{Ae^{-i\delta_0}}{2i} (e^{2i\delta_0} e^{ikr} - e^{-ikr}) \qquad \text{(A-38)}$$

First, we see that this can be done. Second, we find that

$$1 + 2ikf_0 = e^{2i\delta_0}$$

or

$$f_0 = \frac{e^{2i\delta_0} - 1}{2ik} = \frac{e^{i\delta_0} \sin \delta_0}{k} \qquad \text{(A-39)}$$

Hence

$$\frac{d\sigma}{d\Omega} = \frac{\sin^2 \delta_0}{k^2} = \lambdabar^2 \sin^2 \delta_0 \qquad \text{(A-40)}$$

and

$$\sigma = 4\pi \lambdabar^2 \sin^2 \delta_0 \qquad \text{(A-41)}$$

The significance of the phase shift can be appreciated by the following argument. If there were no potential at all, Eq. (A-36) would be the correct Schrödinger equation everywhere, even at $r = 0$; and Eq. (A-37) would be the correct solution. But since $u(0) = 0$ [see Eq. (2-54)], we must set $\delta_0 = 0$ in this case so that

$$u = A \sin kr \qquad \text{(A-42)}$$

is the actual solution. Also $\sigma = 0$, which is obvious, because without scatterer there can be no scattering. The phase shift δ_0 is, therefore, the shift in the phase of the wave function when the potential is "turned on." One can show that if an attractive potential is turned on slowly,[1] δ_0 is always positive at first, and if a repulsive potential is turned on slowly δ_0 is always negative at first. A positive δ_0 means that the wave function is "pulled in," as shown in Fig. A-4.

Scattering length. Another useful parameter, especially in work with very slow neutrons, is called the scattering length.[2] This is defined as the negative of the value of f_0 [Eq. (A-39)] in the limit of zero incident energy or $k \to 0$.

$$a = \lim_{k \to 0} -f_0 \qquad \text{(A-43)}$$

Since f_0 cannot become infinite as $k \to 0$ (otherwise σ would become infinite)

[1] We mean that if the potential is written as $\alpha V(r)$, α is slowly changed from 0 to 1.
[2] This parameter was introduced in Sec. 5-5b.

FIGURE A-4 Definition of the phase shift. Without a scattering potential, the wave function would be of the form $u = A \sin kr$. The presence of a potential $V(r)$ alters the wave function to $u = A \sin (kr + \delta_0)$ far from the scatterer, where $V(r) = 0$. For an attractive potential, the wave function is pulled in and δ_0 is positive.

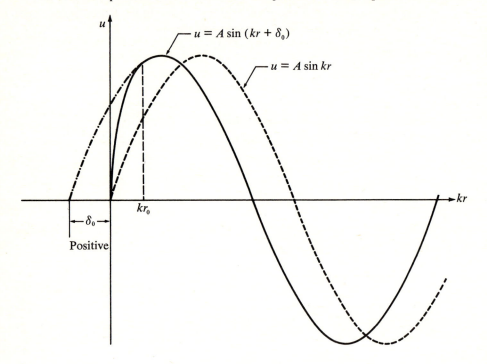

Eq. (A-39) shows that δ_0 must also approach zero and that[1]

$$a = \lim_{k \to 0} - \frac{e^{i\delta_0} \sin \delta_0}{k} = - \frac{\delta_0}{k} \tag{A-44}$$

The cross section (A-41) can then be written

$$\sigma(k \to 0) = 4\pi a^2 \tag{A-45}$$

The following physical significance can be attached to a. At very low energies the wave function (A-37), outside the range of the potential, can be written

$$u = A \sin k(r - a)$$
$$\approx Ak(r - a) \tag{A-46}$$

The scattering length is therefore the extrapolated intercept of the wave function on the r axis, as shown in Fig. A-5. We can also see that if the inner wave

---

[1] The definition (A-43) does not tell whether a will be zero or finite, although physically we expect a finite scattering cross section even at very low energies.

function has a negative slope at $r = r_0$, a is positive.[1] Since a negative slope implies a possible bound state (see Fig. A-1), scattering from a potential giving a bound state produces a positive a. Similarly, if the potential gives only a virtual state, the slope of the inner wave function at $r = r_0$ is positive and a is negative. These cases are shown in Figs. A-5a and b, respectively.

[1] Correctly, we should say that the logarithmic derivative at r_0 determines the sign of a because A could be positive or negative. But $(1/u)(du/dr)$ is independent of A.

FIGURE A-5 Definition of scattering length. At very low energies, the neutron wave function far from the scatterer can be written $u \approx AK(r - a)$. If the potential gives rise to a bound state, a is positive as in (a). If the potential gives rise to an unbound (virtual) state, a is negative as in (b).

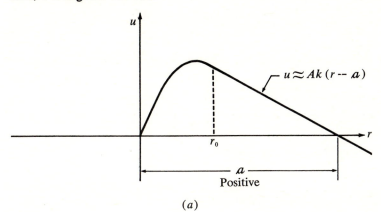

(a)

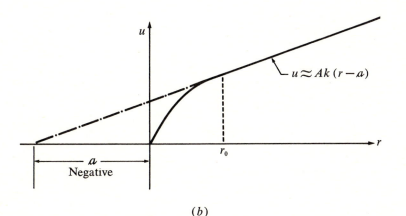

(b)

A-3 NEUTRON-PROTON SCATTERING

The parameterization of the s-wave scattering cross section by the scattering amplitude f_0 or the phase shift δ_0 (or by the scattering length a) is a matter of convenience. We still have to solve the problem in terms of the actual nuclear interaction, but our task is facilitated by the parameterization.

For the n-p scattering problem, we again make the simple assumption of a square well [Eq. (A-1)] for the nuclear potential and solve Eq. (A-3) in the region $r < r_0$. The solution has the mathematical form (A-6)

$$u = C \sin K'r \qquad \text{where} \quad K' = [M(V_0 + T_0)]^{\frac{1}{2}}/\hbar \qquad \text{(A-47)}$$

We recall that T_0 is the c.m. kinetic energy of the incoming particles. The value and derivative of the function (A-47) must be matched at $r = r_0$ to the outer wave function (A-37). Elimination of the constants A and C yields

$$K' \cot K'r_0 = k \cot (kr_0 + \delta_0) \qquad \text{(A-48)}$$

To solve this equation rapidly for δ_0 we make the following temporary simplifications.

1   We assume that the scattering length a [see Eq. (A-44)] is much larger than the range r_0 of the potential, so that, at least at very low neutron energies, kr_0 can be neglected with respect to δ_0.

2   For low incident energies ($T_0 \ll V_0$) we assume

$$K' \approx K \qquad \text{(A-49)}$$

where K is given by Eq. (A-4). The implication is that the shape of the inner wave function is practically independent of the energy E in Eq. (A-3) as long as $|E| \ll V_0$.

Comparing now Eq. (A-48) with Eq. (A-8) we find

$$k \cot \delta_0 \approx -\kappa \qquad \text{(A-50)}$$

where κ is given by Eq. (A-5). Substituting into Eq. (A-40) and noting that $\sin^2 \alpha = 1/(1 + \cot^2 \alpha)$, we finally obtain

$$\frac{d\sigma}{d\Omega} \approx \frac{1}{k^2 + \kappa^2} = \frac{\hbar^2}{M} \frac{1}{T_0 + B} \qquad \text{(A-51)}$$

and

$$\sigma \approx \frac{4\pi\hbar^2}{M} \frac{1}{T_0 + B} = \frac{5.2}{(T_0 + B) \text{ in Mev}} \text{ barn} \qquad \text{(A-52)}$$

For $T_0 \ll B \,(= 2.23 \text{ Mev})$, this gives $\sigma \approx 2.3$ b in strong disagreement with the experimental result of 20.4 b shown in Fig. A-6. (Note that the lab. kinetic energy $T_n = 2T_0$ is plotted along the abscissa of the figure.) At higher energies, Eq. (A-52) approaches the experimental cross section.

FIGURE A-6 Integrated neutron-proton scattering cross section per proton in hydrogen gas. (*a*) Experimental curve. At low energies a rise in the cross section caused by molecular binding effects and thermal motion can be noticed. (*b*) Cross section calculated by Eq. (A-51). The cross section calculated by Eq. (A-55) follows the experimental curve closely. (Experimental cross section from D. J. Hughes and R. B. Schwartz, "Neutron Cross Sections," Brookhaven National Lab., BNL 325, 2d ed., U.S. Government Printing Office, Washington, 1958.)

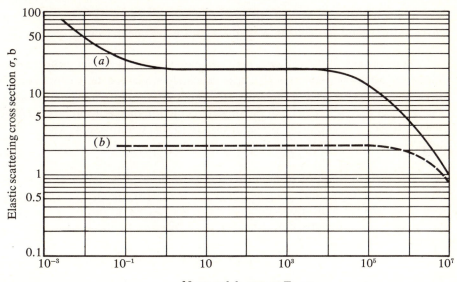

Neutron lab. energy T_n, ev

Wigner (1933)[1] suggested a way to overcome this discrepancy. He noted that in *n-p* scattering, the particles could collide either with total spin 1 (called the triplet state) or with total spin 0 (called the singlet state). If the nuclear interaction should have a different strength or range (or both), Eq. (A-40), and hence Eq. (A-52), would have to be modified.

The probability of colliding in a triplet state is three times that of colliding in a singlet state, because the total spin vector $\mathbf{S}$ can have $2S + 1$ orientations in space. Each orientation is characterized by the magnetic spin quantum number m_S [analogous to m of Eq. (2-44)] which can range from $-S$ to S in integral steps. If $S = 0$, only $m_S = 0$ can occur, but if $S = 1$, $m_S = -1, 0, 1$ are possible. Therefore

$$\frac{d\sigma}{d\Omega} = \frac{3}{4}\frac{\sin^2 \delta_{0t}}{k^2} + \frac{1}{4}\frac{\sin^2 \delta_{0s}}{k^2} \tag{A-53}$$

where the subscripts t and s denote the triplet and singlet contributions respectively.

---

[1] Unpublished, see Bethe and Bacher, 1937.

From the discussion in Sec. A-1, we know that the triplet interaction gives rise to the ground state of the deuteron at an energy $E = -B$. If the singlet interaction produces a state of energy $E = E^*$ (where E^* could be positive or negative), one finds for the integrated cross section

$$\sigma \approx \frac{\pi \hbar^2}{M} \left(\frac{3}{T_0 + B} + \frac{1}{T_0 + |E^*|} \right) \tag{A-55}$$

Comparison with the low-energy n-p cross section yields $|E^*| \approx 70$ kev and gives rather good agreement with the entire experimental cross section (Fig. A-6).

Since the energy E^* is quite small compared to V_0 the singlet wave function inside the nuclear potential is also approximately a one-quarter sine wave, as in the triplet case (Fig. A-1). Therefore we cannot obtain more information about the strength V_{0s} and range r_{0s} of the singlet interaction than is given by Eq. (A-11) for the triplet interaction. In other words, up to this point we only know

$$V_{0t} r_{0t}^2 \approx V_{0s} r_{0s}^2 \approx 1.0 \text{ Mev-barn} \tag{A-56}$$

The zero-range approximation $kr_0 \ll \delta_0$ which we have made [Eq. (A-50)] hides all the information about the range of the nuclear interaction. Had we not made the two simplifications preceding Eq. (A-50), we would have found the same cross section expressions (A-51) and (A-52), but multiplied by a factor[1]

$$(1 + \kappa r_0 + \text{higher-order terms in powers of } \kappa r_0) \tag{A-57}$$

The singlet cross section must also be corrected.[1] Using information from neutron scattering on parahydrogen, discussed in Sec. A-4, and from proton-proton scattering, the various range and strength parameters can be extracted. They are summarized in Sec. A-5 and indicate quite clearly that nuclear forces are spin-dependent, i.e., the interaction is quite different in the singlet and triplet states.

Although we have treated only the square well potential, there is a general method available for extracting range and strength information from nuclear data, which is applicable to potentials of arbitrary radial shape. It is called the *effective range approximation*.[1]

A-4 VIRTUAL STATE OF THE DEUTERON. NEUTRON SCATTERING BY PARAHYDROGEN

Schwinger and Teller (1937) suggested a method of verifying directly that the n-p interaction is indeed spin-dependent. This proposal also allowed them to determine the sign of the singlet excitation energy E^* in Eq. (A-55).

[1] Evans, 1955, chap. 10, sec. 3.

The dimensions of ordinary molecules are of the order of 10^{-8} cm $= 10^5$ F. If neutrons are scattered by molecules, the scattering will take place independently on the various nuclei in each molecule, as long as the neutron (reduced) de Broglie wavelength is much less than the distance between nuclei. This corresponds to neutron energies above about 1 ev (see Table 5-2). But for neutron energies less than approximately 10^{-3} ev, neutron scattering takes place from the entire molecule. We say that the scattering from the nuclei in the molecule is now *coherent*.

From the point of view of scattering theory, the potential V of the molecule in Eq. (A-22) is made up of the individual scattering potentials $V_1 + V_2 + \cdots$ of the nuclei. As long as the neutron wavelength is much larger than the internuclear distance, the scattered wave far from the molecule will move in a radial direction with respect to the center of the molecule. The exact location of the individual nuclei plays no role in this approximation. The total wave function of the scattering problem is, therefore, analogous to Eq. (A-23)

$$\psi = a\{e^{ikz} + [f_1(\theta) + f_2(\theta) + \cdots]r^{-1} e^{ikr}\} \qquad \text{(A-58)}$$

where each scattering amplitude corresponds to a particular nucleus. It then follows from the derivation preceding Eq. (A-20) that

$$\frac{d\sigma}{d\Omega} = |f_1(\theta) + f_2(\theta) + \cdots|^2 \qquad \text{(A-59)}$$

In the case of s-wave scattering, the integrated scattering cross section per molecule (A-35) becomes in obvious notation

$$\sigma = 4\pi |f_{01} + f_{02} + \cdots|^2 \qquad \text{(A-60)}$$

which, at the very low neutron energies implied in our derivation, can also be written in terms of the individual scattering lengths (A-43)

$$\sigma = 4\pi(a_1 + a_2 + \cdots)^2 \qquad \text{(A-61)}$$

Let us apply this to neutron scattering by hydrogen molecules. In a hydrogen molecule, the spins of the two protons can either be aligned (orthohydrogen) or opposed (parahydrogen). At room temperature there is a statistical mixture of these two types of molecules (in the ratio 3:1). At very low temperatures ($<90°$K), hydrogen gas consists only of parahydrogen molecules, because their internal molecular energy is smaller than that of orthohydrogen.

A neutron scattered by a parahydrogen molecule will find itself in the triplet state with respect to one proton and in the singlet state with respect to the other proton. In the scattering cross section (A-61), the triplet and singlet scattering lengths should therefore occur in the ratio 3:1. Taking into account other spin factors, Schwinger and Teller (1937) showed[1]

$$\sigma_{\text{para}} = 6.69 \, (3a_t + a_s)^2$$

Blatt and Weisskopf, 1952, chap. 2, secs. 3.C and D.

Experimentally, $\sigma_{\text{para}} \approx 5$ b, leading to

$$|3a_t + a_s| \approx 8.7 \text{ F} \tag{A-62}$$

Writing the low-energy n-p cross section ($T_n > 1$ ev) at room temperature in terms of the scattering lengths, we find from Eqs. (A-53) and (A-44)

$$\sigma = \pi(3a_t^2 + a_s^2) \tag{A-63}$$

where, by comparison with expression (A-51) or (A-55)

$$a_t \approx \kappa^{-1} = 4.3 \text{ F} \tag{A-64}$$

so that from the experimental value $\sigma = 20.4$ b (Fig. A-6)

$$a_s \approx \pm 24 \text{ F} \tag{A-65}$$

To satisfy Eq. (A-62), we must use the minus sign for a_s. From our discussion at the end of Sec. A-2 and from Fig. A-5, we then see that the singlet interaction must give rise to a virtual state.

By applying Eq. (A-48) at very low energies, we can obtain equations relating the well parameters (r_0 and V_0) to the scattering lengths for the triplet and singlet interactions.[1] This permits us to extract the separate parameters.

A-5 PARAMETERS OF THE TWO-NUCLEON FORCE

An analysis of proton-proton scattering is also very useful, because the low energy p-p scattering can take place only in the singlet s state. A triplet s state for two protons would violate the Pauli exclusion principle.[2] Of course, coulomb scattering has to be considered, but it is modified from Eq. (5-45) because of the identity of the two particles. (After the scattering, it is impossible to distinguish the incident particle from the target nucleus!) Mott (1930) first calculated this quantum-mechanical effect. Coulomb and nuclear scattering interfere coherently. An expression similar to (A-59) must be used for the cross section. The resulting expression is of the form

$$\frac{d\sigma(p, p)}{d\Omega} = \frac{d\sigma_{\text{Mott}}}{d\Omega} + A(\theta, \delta_{0s}) + \frac{\sin^2 \delta_{0s}}{k^2} \tag{A-66}$$

where $A(\theta, \delta_{0s})$ is an interference term. A typical cross section is shown in Fig. A-7. From an analysis of the data, one finds a coulomb-corrected singlet scattering length of -17 F and a singlet range parameter $r_{0s} = 2.7$ F.

Table A-1 summarizes the available information we have discussed. The difference between the singlet and triplet interactions is very marked and confirms the spin dependence of the nuclear force found in complex nuclei. The major difference between the singlet scattering lengths for n-p and for p-p

[1] See Prob. 6-5.

[2] The lowest triplet s state is also absent for the two electrons in a helium atom.

scattering can be attributed to the different magnetic moments of the neutron and proton. A small residual difference has led to the suspicion that the nuclear force might not be completely charge independent. The *n-n* interaction (also possible in the singlet *s* state only, at low energies) can be inferred indirectly from reactions such as $H^2 + \pi^- \to 2n + \gamma$ or $H^2 + n \to 2n + p$. A value of -17 F is consistent with the data and indicates charge symmetry of the nuclear force.

FIGURE A-7   Differential cross section for proton-proton elastic scattering for 2.4-Mev protons. Near a c.m. angle of 90° (lab. angle of 45°), the scattering is mainly caused by nuclear interaction. At forward and backward angles, the coulomb interaction predominates. [By permission from J. D. Jackson and J. M. Blatt, *Rev. Mod. Phys.* **22**:77 (1950).]

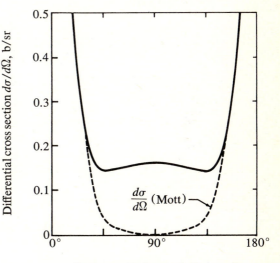

Proton scattering angle Θ_p (c.m.)

TABLE A-1   Square well parameters for the two-nucleon interaction†

Interaction	Scattering length a, F	Range r_0, F	Strength V_0, Mev
Triplet	5.4	2.0	36
Singlet	-23.7‡, -17§	≈2.5	18

† From M. A. Preston, *Physics of the Nucleus*, Addison-Wesley Publishing Co., Inc., Reading, 1962, by permission.
‡ For (*n-p*) interaction.
§ For (*p-p*) interaction.

PHYSICAL PROPERTIES OF THE ELEMENTS[†]

B

Atomic number Z	Element	Atomic or molecular weight $\mathscr{A}$	Density g/cm³ ρ	Nuclei/cm³ ($\times 10^{22}$) n
1	H_2	2.016	8.99×10^{-5}	**
2	He	4.003	17.85×10^{-5}	*
3	Li	6.940	0.534	4.64
4	Be	9.013	1.85	12.37
5	B	10.82	2.34[a]	13.03
6	C	12.01	2.25[b]	11.29
7	N_2	28.02	1.25×10^{-3}	**
8	O_2	32.000	1.43×10^{-3}	**
9	F_2	38.00	1.69×10^{-3}	**
10	Ne	20.18	9.00×10^{-4}	*
11	Na	22.99	0.97	2.54
12	Mg	24.32	1.74	4.31
13	Al	26.98	2.702	6.03
14	Si	28.09	2.329	5.00
15	P	30.98	1.82[c]	3.54

[a]Amorphous   [b]Graphite   [c]Yellow phosphorus, P_4; mol. wt. = 123.92
[d]Rhombic, S_8; mol. wt. = 256.53   [e]Solid, 29.6°C.; melting point = 29.8°C.
[f]Black crystal, As_4; mol. wt. = 299.64   [g]Amorphous, Se_8; mol. wt. = 631.68
[h]Hexagonal   [i]At 22.5°C   [j]Tetragonal (ordinary; β)   [k]Metallic tantalum
*Monoatomic gas; 2.69×10^{19} nuclei/cm³ at NTP   **Diatomic gas; 5.38×10^{19} nuclei/cm³ at NTP

† From J. B. Marion, 1960 Nuclear Data Tables, Part III, Nuclear Data Project, *Natl. Acad. Sci.—Natl. Res. Council, Nucl. Sci. Ser.*, Washington, 1960. Available from U.S. Government Printing Office, Washington, D.C.

Atomic number Z	Element	Atomic or molecular weight $\mathscr{A}$	Density, g/cm^3 ρ	Nuclei/cm^3 ($\times 10^{22}$) n
16	S	32.066	2.07[d]	3.89
17	Cl$_2$	70.91	3.214×10^{-3}	**
18	A	39.944	1.784×10^{-3}	*
19	K	39.10	0.86	1.33
20	Ca	40.08	1.55	2.33
21	Sc	44.96	2.5	3.35
22	Ti	47.90	4.5	5.66
23	V	50.95	5.96	7.05
24	Cr	52.01	7.20	8.34
25	Mn	54.94	7.20	7.90
26	Fe	55.85	7.86	8.48
27	Co	58.94	8.9	9.10
28	Ni	58.71	8.90	9.13
29	Cu	63.54	8.92	8.46
30	Zn	65.38	7.14	6.58
31	Ga	69.72	5.904[e]	5.10
32	Ge	72.60	5.35	4.44
33	As	74.91	5.727[f]	4.61
34	Se	78.96	4.82[g]	3.68
35	Br$_2$	159.83	2.928	2.07
36	Kr	83.80	3.71×10^{-3}	*
37	Rb	85.48	1.532	1.08
38	Sr	87.63	2.6	1.79
39	Y	88.92	5.51	3.73
40	Zr	91.22	6.4	4.23
41	Nb	92.91	8.55	5.54
42	Mo	95.95	10.2	6.40
43	Tc	98	—	—
44	Ru	101.1	12.06[h]	7.19
45	Rh	102.91	12.4	7.26
46	Pd	106.70	11.40[i]	6.44
47	Ag	107.88	10.5	5.86
48	Cd	112.41	8.642	4.63
49	In	114.82	7.30	3.83
50	Sn	118.70	7.28[j]	3.70
51	Sb	121.76	6.684	3.07
52	Te$_2$	255.22	6.25	2.95
53	I$_2$	253.81	4.93	2.34
54	Xe	131.3	5.85×10^{-3}	*
55	Cs	132.91	1.873	0.85
56	Ba	137.36	3.5	1.54
57	La	138.92	6.15	2.67

Atomic number Z	Element	Atomic or molecular weight $\mathscr{A}$	Density, g/cm^3 ρ	Nuclei/cm³ ($\times 10^{22}$) n
58	Ce	140.13	6.7[h]	2.88
59	Pr	140.92	6.5	2.78
60	Nd	144.27	6.9	2.88
61	Pm	145	—	—
62	Sm	150.35	7.7	3.09
63	Eu	152.00	5.22	2.07
64	Gd	157.26	7.95	3.05
65	Td	158.93	8.33	3.16
66	Dy	162.51	8.56	3.17
67	Ho	164.94	8.76	3.20
68	Er	167.20	9.16	3.30
69	Tm	168.94	9.35	3.33
70	Yb	173.04	7.01	2.44
71	Lu	174.99	9.74	3.35
72	Hf	178.60	13.3	4.49
73	Ta	180.95	16.6[k]	5.53
74	W	183.86	19.3	6.32
75	Re	186.22	20.53	6.64
76	Os	190.20	22.48	7.12
77	Ir	192.2	22.42	7.03
78	Pt	195.09	21.45	6.62
79	Au	197.0	19.3	5.90
80	Hg	200.61	13.55	4.07
81	Tl	204.39	11.85	3.49
82	Pb	207.21	11.34	3.30
83	Bi	209.00	9.80	2.83
84	Po	210	9.24	2.65
85	At	211	—	—
86	Rn	222.00	9.73×10^{-3}	*
87	Fr	223	—	—
88	Ra	226.05	5	1.3
89	Ac	227	—	—
90	Th	232.05	11.2	2.94
91	Pa	231	15.4	4.02
92	U	238.07	18.7	4.73
93	Np	237	—	—
94	Pu	239	19.74	4.98

PROPERTIES OF STABLE NUCLIDES†

C

(1) Atomic number; (2) Chemical symbol; (3) Name; (4) Mass number; (5) Neutron number; (6) Mass excess in milli-mass-units based on $C^{12} = 12.000 \ldots u$; (7) Ground-state spin; (8) Ground-state parity: (9) Relative abundance of isotope in element, in percent; (10) Notes. Uncertainties in the last one or two significant figures are given in parentheses.

Z (1)	S (2)	Name (3)	A (4)	N (5)	$10^3(M - A)$ (6)	I (7)	π (8)	Relative abundance (9)	Notes (10)
0	n	Neutron	1*	1	8.66544(43)	$\frac{1}{2}$	+	....	β^-; $t_{\frac{1}{2}} = 12.8$ min.
1	H	Hydrogen	1	0	7.82522(8)	$\frac{1}{2}$	+	99.985	
			2	1	14.10219(11)	1	+	0.015	
2	He	Helium	3	1	16.02994(23)	$\frac{1}{2}$	(+)	$\approx 10^{-4}$	
			4	2	2.60361(37)	0	+	≈ 100	
3	Li	Lithium	6	3	15.1263(10)	1	+	7.35	
			7	4	16.0053(11)	$\frac{3}{2}$	(−)	92.65	
4	Be	Beryllium	9	5	12.1858(9)	$\frac{3}{2}$	−	100	
5	B	Boron	10	5	12.9389(7)	3	+	19.20	
			11	6	9.30509(43)	$\frac{3}{2}$	(−)	80.80	
6	C	Carbon	12	6	0.0000(21)	0	(+)	98.893(5)	
			13	7	3.3543(7)	$\frac{1}{2}$	−	1.107(5)	
7	N	Nitrogen	14	7	3.07438(17)	1	+	99.273(2)	
			15	8	0.1081(9)	$\frac{1}{2}$	−	0.727(2)	
8	O	Oxygen	16	8	−5.08506(28)	0	+	99.5186(7)	
			17	9	−0.8666(9)	$\frac{5}{2}$	+	0.0745(5)	
			18	10	−0.84017(34)	0	+	0.4068(5)	
9	F	Fluorine	19	10	−1.5954(7)	$\frac{1}{2}$	+	100	
10	Ne	Neon	20	10	−7.5596(5)	(0)	(+)	90.920(36)	
			21	11	−6.1508(17)	$\frac{3}{2}$	+	0.258(1)	
			22	12	−8.6155(6)	(0)	(+)	8.822(18)	
11	Na	Sodium	23	12	−10.2274(16)	$\frac{3}{2}$	+	100	
12	Mg	Magnesium	24	12	−14.9554(19)	(0)	(+)	78.6(2)	
			25	13	−14.1603(20)	$\frac{5}{2}$	(+)	10.12(2)	
			26	14	−17.4091(24)	(0)	(+)	11.20(4)	
13	Al	Aluminum	27	14	−18.4651(21)	$\frac{5}{2}$	(+)	100	
14	Si	Silicon	28	14	−23.0729(31)	(0)	(+)	92.17(1)	
			29	15	−23.5092(36)	$\frac{1}{2}$	(+)	4.71(2)	
			30	16	−26.2393(43)	(0)	(+)	3.12(2)	

* Denotes radioactive nucleus. The table includes the neutron and some radioactive nuclei of geological significance.

† From G. L. Trigg, Systematics of Stable Nuclei, in D. W. Gray (ed.), "American Institute of Physics Handbook," 2d ed., McGraw-Hill Book Company, New York, 1963, by permission.

Z (1)	S (2)	Name (3)	A (4)	N (5)	$10^3(M-A)$ (6)	I (7)	π (8)	Relative abundance (9)	Notes (10)
15	P	Phosphorus	31	16	−26.2366(15)	$\frac{1}{2}$	(+)	100	
16	S	Sulfur	32	16	−27.9262(11)	0	+	95.02(30)	
			33	17	−28.5395(30)	$\frac{3}{2}$	+	0.750(15)	
			34	18	−32.1355(31)	0	+	4.215(84)	
			36	20	−32.9095(35)	(0)	(+)	0.017(2)	
17	Cl	Chlorine	35	18	−31.1455(28)	$\frac{3}{2}$	+	75.529(24)	
			37	20	−34.1041(22)	$\frac{3}{2}$	+	24.471(24)	
18	Ar (A)	Argon	36	18	−32.4519(34)	(0)	(+)	0.337(1)	
			38	20	−37.2755(24)	(0)	(+)	0.063(1)	
			40	22	−37.6162(8)	(0)	(+)	99.600(1)	
19	K	Potassium	39	20	−36.2860(30)	$\frac{3}{2}$	(+)	93.126(5)	
			40*	21	−35.9921(36)	4	(−)	0.0112(5)	EC(12.4%), β^-(87.6%); $t_{\frac{1}{2}} = 1.28 \times 10^9$ years.
			41	22	−38.1649(46)	$\frac{3}{2}$	+	6.862(5)	
20	Ca	Calcium	40	20	−37.4108(37)	(0)	(+)	96.92(3)	
			42	22	−41.3723(44)	(0)	(+)	0.64(1)	
			43	23	−41.2200(48)	$\frac{7}{2}$	(−)	0.132(4)	
			44	24	−44.5103(48)	(0)	(+)	2.13(4)	
			46	26	−46.3112(41)	(0)	(+)	0.0032	
			48	28	−47.481(15)	(0)	(+)	0.0179(7)	
21	Sc	Scandium	45	24	−44.0811(42)	$\frac{7}{2}$	(−)	100	
22	Ti	Titanium	46	24	−47.3666(37)	(0)	(+)	7.99(2)	
			47	25	−48.242(8)	$\frac{5}{2}$	(−)	7.32(2)	
			48	26	−52.0522(36)	(0)	(+)	73.99(7)	
			49	27	−52.1334(35)	$\frac{7}{2}$	(−)	5.46(2)	
			50	28	−55.2109(48)	(0)	(+)	5.25(5)	
23	V	Vanadium	50*	27	−52.8354(40)	6	(+)	0.24(1)	EC; $t_{\frac{1}{2}} = 4 \times 10^{14}$ years.
			51	28	−56.0221(42)	$\frac{7}{2}$	(−)	99.76(1)	
24	Cr	Chromium	50	26	−53.9493(45)	(0)	(+)	4.31(4)	
			52	28	−59.4863(36)	(0)	(+)	83.76(14)	
			53	29	−59.3489(37)	$\frac{3}{2}$	(−)	9.55(9)	
			54	30	−61.1206(48)	(0)	(+)	2.38(2)	
25	Mn	Manganese	55	30	−61.9464(41)	$\frac{5}{2}$	(−)	100	
26	Fe	Iron	54	28	−60.379(6)	(0)	(+)	5.81(1)	
			56	30	−65.068(6)	(0)	(+)	91.64(2)	
			57	31	−64.606(6)	$\frac{1}{2}$	(−)	2.21(1)	
			58	32	−66.728(7)	(0)	(+)	0.34(1)	
27	Co	Cobalt	59	32	−66.8109(46)	$\frac{7}{2}$	(−)	100	
28	Ni	Nickel	58	30	−64.658(6)	(0)	(+)	67.76(22)	
			60	32	−69.217(6)	(0)	(+)	26.16(66)	
			61	33	−68.951(9)	($\frac{3}{2}$)	(−)	1.25(3)	
			62	34	−71.655(7)	(0)	(+)	3.66(1)	
			64	36	−72.041(6)	(0)	(+)	1.16(20)	
29	Cu	Copper	63	34	−70.406(6)	$\frac{3}{2}$	—	69.12(5)	
			65	36	−72.214(6)	$\frac{3}{2}$	—	30.88(5)	
30	Zn	Zinc	64	34	−70.855(6)	(0)	(+)	48.89	
			66	36	−73.952(10)	(0)	(+)	27.81	
			67	37	−72.851(11)	$\frac{5}{2}$	—	4.11	
			68	38	−75.135(9)	(0)	(+)	18.56	
			70	40	−74.652(16)	(0)	(+)	0.62	
31	Ga	Gallium	69	38	−74.318(28)	$\frac{3}{2}$	—	60.22(16)	
			71	40	−75.16(5)	$\frac{3}{2}$	—	39.78(16)	
32	Ge	Germanium	70	38	−75.723(20)	(0)	(+)	20.52(17)	
			72	40	−78.26(5)	(0)	(+)	27.43(21)	
			73	41	−76.64(7)	$\frac{9}{2}$	+	7.76(8)	
			74	42	−78.85(6)	(0)	(+)	36.54(23)	
			76	44	−78.64(9)	(0)	(+)	7.76(8)	
33	As	Arsenic	75	42	−78.42(5)	$\frac{3}{2}$	—	100	

Z (1)	S (2)	Name (3)	A (4)	N (5)	$10^3(M-A)$ (6)	I (7)	π (8)	Relative abundance (9)	Notes (10)
34	Se	Selenium	74	40	−77.55(6)	0	(+)	0.87(1)	
			76	42	−80.771(48)	(0)	(+)	9.02(7)	
			77	43	−80.066(48)	½	−	7.58(7)	
			78	44	−82.652(48)	0	(+)	23.52(2)	
			80	46	−83.488(17)	0	(+)	49.82(20)	
			82	48	−83.34(7)	(0)	(+)	9.19(20)	
35	Br	Bromine	79	44	−81.652(19)	3/2	−	50.537(10)	
			81	46	−83.656(37)	3/2	−	49.463(10)	
36	Kr	Krypton	78	42	−79.632(5)	(0)	(+)	0.354(2)	
			80	44	−83.612(13)	(0)	(+)	2.27(1)	
			82	46	−86.517(8)	(0)	(+)	11.56(2)	
			83	47	−85.869(8)	9/2	+	11.55(2)	
			84	48	−88.496(5)	(0)	(+)	56.90(12)	
			86	50	−89.383(8)	(0)	(+)	17.37(3)	
37	Rb	Rubidium	85	48	−88.29(6)	5/2	−	72.15(5)	
			87*	50	−90.82(8)	3/2	−	27.85(5)	β^-; $t_{\frac{1}{2}} = 5.0 \times 10^{10}$ years.
38	Sr	Strontium	84	46	−86.624(11)	(0)	(+)	0.55(1)	
			86	48	−90.74(8)	(0)	(+)	9.75(4)	
			87	49	−91.11(8)	9/2	+	6.96(1)	
			88	50	−94.39(9)	(0)	(+)	82.74(6)	
39	Y	Yttrium	89	50	−94.57(9)	½	−	100	
40	Zr	Zirconium	90	50	−95.68(9)	(0)	(+)	51.46	
			91	51	−94.75(10)	5/2	+	11.23	
			92	52	−95.41(11)	(0)	(+)	17.11	
			94	54	−93.86(36)	(0)	(+)	17.40	
			96	56	−91.8(8)	(0)	(+)	2.80	
41	Nb	Niobium	93	52	−93.98(11)	9/2	+	100	Formerly known as columbium, chemical symbol Cb.
42	Mo	Molybdenum	92	50	−93.71(13)	(0)	(+)	15.86(16)	
			94	52	−95.26(13)	(0)	(+)	9.12(9)	
			95	53	−94.28(36)	5/2	...	15.70(16)	
			96	54	−95.45(36)	(0)	(+)	16.50(17)	
			97	55	−94.25(40)	5/2	...	9.45(10)	
			98	56	−94.49(41)	(0)	(+)	23.75(8)	
			100	58	−92.43(49)	(0)	(+)	9.62(10)	
44	Ru	Ruthenium	96	52	−92.4(7)	(0)	(+)	5.57(8)	
			98	54	−94.5(8)	(0)	(+)	1.86(4)	
			99	55	−93.92(49)	5/2	(+)	12.7(1)	
			100	56	−95.782(5)	(0)	(+)	12.6(1)	‡
			101	57	−94.423(3)	5/2	(+)	17.1(1)	‡
			102	58	−96.28(20)	(0)	(+)	31.6(2)	
			104	60	−94.47(40)	(0)	(+)	18.5(1)	
45	Rh	Rhodium	103	58	−95.20(20)	½	−	100	
46	Pd	Palladium	102	56	−95.06(19)	(0)	(+)	0.80(1)	
			104	58	−96.44(20)	(0)	(+)	9.3(1)	
			105	59	−95.36(27)	5/2	+	22.6(2)	
			106	60	−96.80(12)	(0)	(+)	27.2(3)	
			108	62	−96.08(12)	(0)	(+)	26.8(3)	
			110	64	−95.50(32)	(0)	(+)	13.5(1)	
47	Ag	Silver	107	60	−95.03(11)	½	−	51.35(7)	
			109	62	−95.30(11)	½	−	48.65(7)	

‡ From J. H. E. Mattauch, W. Thiele, and A. H. Wapstra, 1964, Atomic Mass Table, *Nuclear Phys.*, **67**:1 (1965).

Z (1)	S (2)	Name (3)	A (4)	N (5)	$10^3(M-A)$ (6)	I (7)	π (8)	Relative abundance (9)	Notes (10)
48	Cd	Cadmium	106	58	−94.05(37)	(0)	(+)	1.215	
			108	60	−96.00(12)	(0)	(+)	0.875	
			110	62	−97.03(11)	(0)	(+)	12.39	
			111	63	−95.85(19)	½	+	12.75	
			112	64	−97.16(11)	(0)	(+)	24.07	
			113	65	−95.39(10)	½	+	12.26	
			114	66	−96.43(10)	(0)	(+)	28.86	
			116	68	−94.99(32)	(0)	(+)	7.58	
49	In	Indium	113	64	−95.72(10)	9/2	+	4.26(5)	Stability against EC not certain; $t_{\frac{1}{2}} \geq 10^{14}$ years. β^-; $t_{\frac{1}{2}} = 6 \times 10^{14}$ years.
			115*	66	−95.93(10)	9/2	+	95.74(5)	
50	Sn	Tin	112	62	−95.06(11)	(0)	(+)	0.90(1)	
			114	64	−97.04(10)	(0)	(+)	0.61(1)	
			115	65	−96.47(11)	½	+	0.35(1)	
			116	66	−97.89(19)	(0)	(+)	14.07(8)	
			117	67	−96.94(19)	½	+	7.54(3)	
			118	68	−98.21(19)	(0)	(+)	23.98(3)	
			119	69	−96.61(20)	½	+	8.62(1)	
			120	70	−97.87(14)	(0)	(+)	33.03(12)	
			122	72	−96.59(14)	(0)	(+)	4.78(1)	
			124	74	−94.76(13)	(0)	(+)	6.11(1)	
51	Sb	Antimony	121	70	−96.25(14)	5/2	+	57.25(3)	
			123	72	−95.85(14)	7/2	+	42.75(3)	
52	Te	Tellurium	120	68	−95.49(40)	(0)	(+)	0.091(1)	
			122	70	−97.00(13)	(0)	(+)	2.49(2)	
			123	71	−95.82(13)	½	+	0.89(2)	EC indicated by mass, but uncertain; $t_{\frac{1}{2}} \geq 10^{12}$ years.
			124	72	−97.24(13)	(0)	(+)	4.63(5)	
			125	73	−95.58(13)	½	+	7.01(1)	
			126	74	−96.758(37)	(0)	(+)	18.72(4)	
			128	76	−95.29(14)	(0)	(+)	31.72(1)	
			130	78	−93.30(14)	(0)	(+)	34.46(9)	
53	I	Iodine	127	74	−95.648(23)	5/2	+	100	
54	Xe	Xenon	124	70	−93.88(16)	(0)	(+)	0.09614(36)	
			126	72	−95.831(32)	(0)	(+)	0.08956(36)	
			128	74	−96.462(10)	(0)	(+)	1.919(4)	
			129	75	−95.216(10)	½	+	26.44(7)	
			130	76	−96.490(9)	(0)	(+)	4.074(10)	
			131	77	−94.913(7)	3/2	+	21.18(5)	
			132	78	−95.838(8)	(0)	(+)	26.89(6)	
			134	80	−94.602(8)	(0)	(+)	10.44(2)	
			136	82	−92.779(10)	(0)	(+)	8.869(9)	
55	Cs	Cesium	133	78	−94.91(15)	7/2	+	100	
56	Ba	Barium	130	74	−93.753(24)	(0)	(+)	0.13(2)	
			132	76	−94.88(32)	(0)	(+)	0.19(2)	
			134	78	−95.69(15)	(0)	(+)	2.66(5)	
			135	79	−94.43(26)	3/2	(+)	6.73(12)	
			136	80	−95.64(14)	(0)	(+)	8.07(10)	
			137	81	−94.44(13)	3/2	+	11.87(25)	
			138	82	−94.99(8)	(0)	(+)	70.41(35)	
57	La	Lanthanum	138*	81	−93.19(8)	5	−	0.089(1.5)	EC 70%, β^- 30%; $t_{\frac{1}{2}} = 1.0 \times 10^{11}$ years.
			139	82	−93.94(8)	7/2	(+)	99.911(1.5)	
58	Ce	Cerium	136	78	−92.9(5)	(0)	(+)	0.193(5)	
			138	80	−94.28(8)	(0)	(+)	0.250(5)	
			140	82	−94.72(5)	(0)	(+)	88.48(10)	
			142	84	−90.96(8)	(0)	(+)	11.07(10)	
59	Pr	Praseodymium	141	82	−92.610(46)	5/2	+	100	

Z (1)	S (2)	Name (3)	A (4)	N (5)	$10^3(M-A)$ (6)	I (7)	π (8)	Relative abundance (9)	Notes (10)
60	Nd	Neodymium	142	82	−92.522(47)	(0)	(+)	27.09(3)	
			143	83	−90.38(5)	$\frac{7}{2}$	(−)	12.14(2)	
			144*	84	−90.10(5)	(0)	(+)	23.83(3)	α; $t_{\frac{1}{2}} = 5 \times 10^{15}$ years.
			145	85	−87.84(15)	$\frac{7}{2}$	(−)	8.29(1)	
			146	86	−87.31(15)	(0)	(+)	17.26(2)	
			148	88	−83.52(16)	(0)	(+)	5.74(2)	
			150	90	−79.29(15)	(0)	(+)	5.63(2)	
62	Sm	Samarium	144	82	−88.35(24)	(0)	(+)	3.02(2)	
			147*	85	−85.38(5)	$\frac{7}{2}$	(−)	14.87(4)	α; $t_{\frac{1}{2}} = 1.3 \times 10^{11}$ years.
			148	86	−85.44(13)	(0)	(+)	11.22(3)	
			149	87	−83.07(13)	$\frac{7}{2}$	(−)	13.82(4)	
			150	88	−82.99(13)	(0)	(+)	7.40(2)	
			152	90	−80.65(33)	(0)	(+)	26.80(5)	
			154	92	−78.33(30)	(0)	(+)	22.88(6)	
63	Eu	Europium	151	88	−80.37(18)	$\frac{5}{2}$	(+)	47.86(8)	
			153	90	−79.28(35)	$\frac{5}{2}$	+	52.14(8)	
64	Gd	Gadolinium	152	88	−80.60(33)	(0)	(+)	0.205(1)	
			154	90	−79.77(28)	(0)	(+)	2.23(3)	
			155	91	−77.99(26)	$\frac{3}{2}$	(−)	15.1(1.5)	
			156	92	−77.76(26)	(0)	(+)	20.6(2)	
			157	93	−75.96(27)	$\frac{3}{2}$	(−)	15.7(1.6)	
			158	94	−75.81(27)	(0)	(+)	24.5(2.5)	
			160	96	−72.7(5)	(0)	(+)	21.6(2)	
65	Tb	Terbium	159	94	−75.7(11)	$\frac{3}{2}$	(+)	100	
66	Dy	Dysprosium	156	90	−76.07(18)	(0)	(+)	0.0524(5)	‡
			158	92	−75.55(3)	(0)	(+)	0.0902(9)	‡
			160	94	−76.0(10)	(0)	(+)	2.294(11)	
			161	95	−74.2(10)	$\frac{5}{2}$	(+)	18.88(9)	
			162	96	−74.3(10)	(0)	(+)	25.53(13)	
			163	97	−72.4(10)	$\frac{5}{2}$	(−)	24.97(12)	
			164	98	−71.9(10)	(0)	(+)	28.18(12)	
67	Ho	Holmium	165	98	−70.4(7)	$\frac{7}{2}$	(−)	100	
68	Er	Erbium	162	94	−71.26(9)	(0)	(+)	0.136(3)	‡
			164	96	−71.4(7)	(0)	(+)	1.56(3)	
			166	98	−68.6(6)	(0)	(+)	33.41(3)	
			167	99	−67.9(6)	$\frac{7}{2}$	(+)	22.94(2)	
			168	100	−68.6(6)	(0)	(+)	27.07(3)	
			170	102	−64.9(21)	(0)	(+)	14.88(2)	
69	Tm	Thulium	169	100	−65.755(34)	$\frac{1}{2}$	(+)	100	‡
70	Yb	Ytterbium	168	98	−65.84(16)	(0)	(+)	0.135(2)	‡
			170	100	−64.98(6)	(0)	(+)	3.14(4)	‡
			171	101	−63.57(7)	$\frac{1}{2}$	(−)	14.4(1.5)	‡
			172	102	−63.64(7)	(0)	(+)	21.9(2.5)	‡
			173	103	−61.94(7)	$\frac{5}{2}$	(−)	16.2(2)	‡
			174	104	−61.26(6)	(0)	(+)	31.6(4)	‡
			176	106	−57.32(7)	(0)	(+)	12.6(1.5)	‡
71	Lu	Lutecium	175	104	−59.36(6)	$\frac{7}{2}$	(+)	97.412(13)	‡
			176*	105	−58.56(46)	...	...	2.588(13)	β^-; EC questionable. $t_{\frac{1}{2}} = 2.4 \times 10^{10}$ years.
72	Hf	Hafnium	174	102	−59.64(7)	(0)	(+)	0.163(2)	‡
			176	104	−59.66(46)	(0)	(+)	5.21(2)	
			177	105	−58.08(46)	$\frac{7}{2}$	(−)	18.56(6)	
			178	106	−57.51(44)	(0)	(+)	27.10(10)	
			179	107	−55.58(44)	$\frac{9}{2}$	(+)	13.75(5)	
			180	108	−54.88(43)	(0)	(+)	35.22(10)	
73	Ta	Tantalum	180	107	−54.28(38)	...	...	0.0123(3)	Mode of decay not established; $t_{\frac{1}{2}} \geq 1.7(3) \times 10^{13}$ years.
			181	108	−53.82(38)	$\frac{7}{2}$	+	99.9877(3)	

Z (1)	S (2)	Name (3)	A (4)	N (5)	$10^3(M-A)$ (6)	I (7)	π (8)	Relative abundance (9)	Notes (10)
74	W	Tungsten	180	106	−55.03(38)	(0)	(+)	0.126(6)	The name "wolfram" is gradually becoming common.
			182	108	−53.53(38)	(0)	(+)	26.31(3)	
			183	109	−51.51(38)	$\frac{1}{2}$	(−)	14.28(1)	
			184	110	−50.85(39)	(0)	(+)	30.64(3)	
			186	112	−48.6(16)	(0)	(+)	28.64(1)	
75	Re	Rhenium	185	110	−49.9(16)	$\frac{5}{2}$	+	37.07(6)	
			187*	112	−45.02(34)	$\frac{5}{2}$	+	62.93(6)	$t_\frac{1}{2}$ variously reported as $\leq 10^{11}$ and $> 10^{16}$ years. ‡
76	Os	Osmium	184	108	−47.25(7)	(0)	(+)	0.018(2)	
			186	110	−47.06(35)	(0)	(+)	1.59(5)	
			187	111	−45.03(34)	$\frac{1}{2}$	(−)	1.64(5)	
			188	112	−45.02(30)	(0)	(+)	13.3(2)	
			189	113	−42.78(33)	$\frac{3}{2}$	(−)	16.1(2)	
			190	114	−42.58(36)	(0)	(+)	26.4(3)	
			192	116	−39.49(34)	(0)	(+)	41.0(2)	
77	Ir	Iridium	191	114	−40.10(29)	$\frac{3}{2}$	+	38.5	
			193	116	−37.66(29)	$\frac{3}{2}$	+	61.5	
78	Pt	Platinum	190*	112	−40.83(41)	(0)	(+)	0.0127(5)	α; $t_\frac{1}{2} = 5.9 \times 10^{11}$ years.
			192*	114	−39.53(29)	(0)	(+)	0.78(1)	α; $t_\frac{1}{2} \approx 10^{15}$ years.
			194	116	−37.57(24)	(0)	(+)	32.9(1)	
			195	117	−35.54(24)	$\frac{1}{2}$	−	33.8(1)	
			196	118	−35.38(24)	(0)	(+)	25.2(1)	
			198	120	−32.47(31)	(0)	(+)	7.19(4)	
79	Au	Gold	197	118	−33.448(16)	$\frac{3}{2}$	+	100	
80	Hg	Mercury	196	116	−34.181(18)	0	(+)	0.146(5)	
			198	118	−33.231(15)	(0)	(+)	10.02(1)	
			199	119	−31.744(20)	$\frac{1}{2}$	−	16.84(4)	
			200	120	−31.656(14)	(0)	(+)	23.13(8)	
			201	121	−29.685(18)	$\frac{3}{2}$	(−)	13.22(5)	
			202	122	−29.370(23)	(0)	(+)	29.80(3)	
			204	124	−26.518(19)	(0)	(+)	6.85(1)	
81	Tl	Thallium	203	122	−27.669(40)	$\frac{1}{2}$	+	29.50	
			205	124	−25.538(27)	$\frac{1}{2}$	(+)	70.50	
82	Pb	Lead	204	122	−26.931(24)	(0)	(+)	1.4	
			206	124	−25.541(12)	(0)	(+)	25	
			207	125	−24.102(12)	$\frac{1}{2}$	−	22	
			208	126	−23.356(12)	(0)	(+)	52	
83	Bi	Bismuth	209*	126	−19.583(27)	$\frac{9}{2}$	(−)	100	Activity disputed: α; $t_\frac{1}{2} = 2 \times 10^{17}$ years.
90	Th	Thorium	232*	142	38.211(42)	(0)	(+)	100	α; $t_\frac{1}{2} = 1.39 \times 10^{10}$ years.
92	U	Uranium	234*	142	40.90(6)	(0)	(+)	0.0057(2)	α; $t_\frac{1}{2} = 2.48 \times 10^{5}$ years.
			235*	143	43.933(43)	$\frac{7}{2}$	(−)	0.7204(7)	α; $t_\frac{1}{2} = 7.1 \times 10^{8}$ years.
			238*	146	50.76(8)	(0)	(+)	99.2739(7)	α; $t_\frac{1}{2} = 4.51 \times 10^{9}$ years.

VALUES OF PHYSICAL CONSTANTS AND CONVERSION FACTORS† D

General physical constants. Least-squares adjusted output values of 1963. The digits in parentheses following each quoted value represent the standard deviation error in the final digits of the quoted value as computed on the criterion of internal consistency. The unified scale of atomic weights is used throughout ($^{12}C = 12$). C = coulomb; J = joule; N = newton; u = mass unit.

Constant	Value	Unit mks	Unit cgs
Speed of light in vacuum c	2.997925(1)	$\times 10^8$ m s^{-1}	$\times 10^{10}$ cm s^{-1}
Elementary charge e	1.60210(2)	10^{-19} C	10^{-20} emu
	4.80298(7)		10^{-10} esu
Avogadro's number $\mathcal{N}$	6.02252(9)	10^{26} kmole^{-1}	10^{23} mole^{-1}
Mass unit	1.66043(2)	10^{-27} kg	10^{-24} g
Electron rest mass m_0	9.10908(13)	10^{-31} kg	10^{-28} g
	5.48597(3)	10^{-4} u	10^{-4} u
Proton rest mass M_P	1.67252(3)	10^{-27} kg	10^{-24} g
	1.00727663(8)	u	u
Neutron rest mass M_n	1.67482(3)	10^{-27} kg	10^{-24} g
	1.0086654(4)	u	u
Faraday constant $\mathcal{N}e$	9.64870(5)	10^4 C mole^{-1}	10^3 emu
	2.89261(2)		10^{14} esu
Planck constant h, $\hbar = h/2\pi$	6.62559(16)	10^{-34} J s	10^{-27} erg s
	1.054494(25)	10^{-34} J s	10^{-27} erg s
Charge-to-mass ratio for electron e/m_0	1.758796(6)	10^{11} C kg^{-1}	10^7 emu
	5.27274(2)		10^{17} esu
Rydberg constant $2\pi^2 m_0 e^4/h^3 c$	1.0973731(1)	10^7 m^{-1}	10^5 cm^{-1}
Bohr radius $\hbar^2/m_0 e^2$	5.29167(2)	10^{-11} m	10^{-9} cm
Compton wavelength of electron $h/m_0 c$, $\hbar/m_0 c$	2.42621(2)	10^{-12} m	10^{-10} cm
	3.86144(3)	10^{-13} m	10^{-11} cm
Compton wavelength of proton $h/M_p c$, $\hbar/M_p c$	1.321398(13)	10^{-15} m	10^{-13} cm
	2.10307(2)	10^{-16} m	10^{-14} cm

† From E. R. Cohen and J. W. M. DuMond, *Rev. Mod. Phys.*, **37**: 537 (1965), by permission.

Conversion factor	*Value*
1 electron volt	$1.60210(2) \times 10^{-19}$ J
	$1.60210(2) \times 10^{-12}$ erg
	$8065.73(8)$ cm^{-1}
	$2.41804(2) \times 10^{14}$ s^{-1}
$E_r \lambda_r$	$12398.10(13) \times 10^{-8}$ ev cm
1 u	$931.478(5)$ Mev
Proton mass $M_p c^2$	$938.256(5)$ Mev
Neutron mass $M_n c^2$	$939.550(5)$ Mev
Electron mass $m_0 c^2$	$511006(2)$ ev
Rydberg $2\pi^2 m_0 e^4/h^2$	$2.17971(5) \times 10^{-11}$ erg
	$13.60535(13)$ ev
Gas constant	8.31434×10^7 erg mole^{-1} deg^{-1}
	0.082053 liter atm mole^{-1} deg^{-1}
	82.055 cm^3 atm mole^{-1} deg^{-1}
	1.9872 cal$_{\text{th}}$ mole^{-1} deg^{-1}
Standard volume of ideal gas at NTP	22413.6 cm^3 mole^{-1}
$\dfrac{\text{Mass on physical scale (O}^{16}=16)}{\text{Mass on unified scale (C}^{12}=12)}$	$1.000317917(17)$
$\dfrac{\text{Mass on chemical scale (O}=16)}{\text{Mass on unified scale (C}^{12}=12)}$	$1.000043(5)$

BIBLIOGRAPHY

Each reference gives at least the original paper and mentions in brackets the section in this book where it is quoted. Sometimes the reader is also referred to review articles. More references and reading selections suitable for our level of presentation can be found in J. G. Cunninghame, "Introduction to the Atomic Nucleus," Elsevier Publishing Company, New York, 1964.

Alvarez, L.: *Phys. Rev.*, **52**:134 (1937). [Sec. 4-6f]

Anderson, C. D.: *Science*, **76**:238 (1932); *Phys. Rev.*, **43**:491 (1933). [Sec. 3-4d]

——— and S. H. Neddermeyer: *Phys. Rev.*, **50**:263 (1936); S. H. Neddermeyer and C. D. Anderson: *Phys. Rev.*, **51**:884 (1937). [Sec. 1-1]

Aston, F. W.: *Phil. Mag.*, **38**:709 (1919); "Mass Spectra and Isotopes," Edward Arnold (Publishers) Ltd., London, 1933. [Sec. 1-2a]

Barkla, C. G.:   *Phil. Mag.*, **21**:648 (1911). [Sec. 1-2a]

Bartlett, J. H.:   *Nature*, **130**:165 (1932). [Sec. 2-5]

Bequerel, H.:   *Compt. Rend.*, **122**:420, 501 (1896). [Secs. 1-1, 4-5]

Bethe, H. A.:   *Phys. Rev.*, **57**:1125 (1940). [Sec. 5-1]

———— and R. F. Bacher: *Rev. Mod. Phys.*, **8**:117 (1937). [Sec. A-3]

Blackett, P. M. S., and D. S. Lees: *Proc. Roy. Soc.* (*London*) **A136**:325 (1932). [Sec. 5-2a]

Blatt, J. B., and V. F. Weisskopf: "Theoretical Nuclear Physics," John Wiley & Sons, Inc., New York, 1952. [Secs. 2-3b, 2-6, 4-4d, 5-5, 5-5a, A-4]

Bloch, F.:   *Ann. Physik*, **16**:285 (1933); *Z. Physik*, **81**:363 (1933). [Sec. 3-2]

Bohr, N.:   *Phil. Mag.*, **26**:1, 476, 857 (1913). [Secs. 1-1, 2-2a]

————:   *Z. Physik*, **13**:117 (1923). [Sec. 2-2f]

————:   *Nature*, **137**:344 (1936). [Sec. 5-1]

———— and J. A. Wheeler: *Phys. Rev.*, **56**:426 (1939). [Sec. 5-7b]

Born, M.:   "Problems of Atomic Dynamics," M.I.T., Cambridge, Mass. 1936; *Z. Physik*, **37**:863 (1926); **38**:803 (1926); **40**:167 (1927). [Secs. 2-2b, 2-2c]

Brink, D. M.:   "Nuclear Forces," Pergamon Press, London, 1965. [Secs. 6-2, A-2]

Brueckner, K. A., A. M. Lockett, and M. Rotenberg: *Phys. Rev.*, **121**:255 (1961). [Sec. 2-4]

Burcham, W. E.:   "Nuclear Physics," McGraw-Hill Book Company, New York, 1963. [Secs. 1-1, 1-2c, 2-6, 3-1, 3-2, 3-3b, 3-4a, 3-4b, 3-6, 4-6d, 4-6h, 5-4d, 5-5a]

Chadwick, J.:   *Verhandl. Deut. Physik. Ges.*, **16**:383 (1914). [Sec. 4-6]

————:   *Nature*, **129**:312 (1932); *Proc. Roy. Soc.* (*London*) **A136**:692 (1932). [Secs. 1-1, 1-2a]

Compton, A. H.:   *Phys. Rev.*, **21**:483, 715 (1923); **22**:409 (1923). [Sec. 2-2a]

Cooper, J. A., J. M. Hollander, and J. O. Rasmussen: *Phys. Rev. Letters*, **15**:680 (1965). [Sec. 4-2a]

Curie, I., and F. Joliot: *Compt. Rend.*, **198**:254,559 (1934). [Sec. 4-6]

Curie, P. and M.:   *Compt. Rend.*, **127**:175,1215 (1898). [Sec. 1-1]

Davis, R.:   *Phys. Rev.*, **97**:766 (1955). [Sec. 4-6g]

Davisson, C., and L. H. Germer: *Phys. Rev.*, **30**:705 (1927); *Proc. Natl. Acad. Sci. US*, **14**:317, 619 (1928). [Sec. 2-2]

Dearnaley, G., and D. C. Northrup: "Semiconductor Counters for Nuclear Radiations," John Wiley & Sons, Inc., New York, 1963. [Sec. 3-6]

DeBenedetti, S.:   "Nuclear Interactions," John Wiley & Sons, Inc., New York, 1964. [Sec. 4-4d]

de Broglie, L.:   *Phil. Mag.*, **47**:446 (1924); *Ann. Phys.* (*Paris*), **10**:3, 22 (1925). [Sec. 2-2a]

Deutsch, M.:   *Phys. Rev.*, **82**:455 (1951); see also M. Deutsch and S. Berko: Positron Annihilation and Positronium, in K. Siegbahn (ed.), "Alpha-, Beta-, and

Gamma-Ray Spectroscopy," chap. 26, North-Holland Publishing Company, Amsterdam, 1965. [Sec. 3-5]

Dirac, P. A. M.: *Proc. Roy. Soc. (London)*, **A117**:610 (1928); **A118**:351 (1928). [Sec. 2-2a]

Einstein, A.: *Ann. Physik*, **17**:132 (1905). [Sec. 2-2a]

Elsasser, W. M.: *J. Phys. Radium*, **4**:549 1(933); **5**:389, 635 (1934). [Sec. 2-5]

Evans, R. D.: "The Atomic Nucleus," McGraw-Hill Book Company, New York, 1955. [Secs. 2-2g, 2-3a, 3-1, 3-2, 3-4a, 3-4d, 3-6, 4-2a, 4-2d, 4-5b, 4-5c, 5-2a, 5-7b, A-3]

Fermi, E.: *Z. Physik*, **88**:161 (1934). [Sec., 4-6c]

Fernbach, S., R. Serber, and T. B. Taylor: *Phys. Rev.*, **75**:1352 (1949). [Sec. 5-1]

Feshbach, H.: *Ann. Phys. (NY)*, **5**:357 (1958); **19**:287 (1960). [Sec. 5-1]

———, C. E. Porter, and V. F. Weisskopf: *Phys. Rev.*, **96**:448 (1954). [Sec. 5-1]

Foldy, L.: *Phys. Today*, **18**:26 (1965). [Sec. 6-2]

Franklin, P.: "A Treatise on Advanced Calculus," John Wiley & Sons, Inc., New York, 1940. [Sec. 4-3]

Gamow, G.: *Z. Physik*, **51**:204 (1928). [Secs. 1-1, 4-5b]

Geiger, H., and E. Marsden: *Proc. Roy. Soc. (London)* **A82**:495 (1909). [Secs. 1-1, 1-2b]

——— and J. M. Nuttall: *Phil. Mag.*, **22**:613 (1911); **23**:439 (1912). [Sec. 4-5b]

Goldhaber, M., L. Grodzins, and A. W. Sunyar: *Phys. Rev.*, **109**:1015 (1958); see also Helicity of the Neutrino, in K. Siegbahn (ed.), "Alpha-, Beta-, and Gamma-Ray Spectroscopy," chap. 24, sec. E, North-Holland Publishing Company, Amsterdam, 1965. [Sec. 4-6g]

Gomez, L. C., J. D. Walecka, and V. F. Weisskopf: *Ann. Phys. (NY)*, **3**:241 (1958). [Sec. 2-3b]

Groshev, L. V., V. N. Lutsenko, A. M. Demidov, and V. I. Pelekov: "Atlas of Gamma-Ray Spectra from Radiative Capture of Thermal Neutrons," Pergamon Press, New York, 1959. [Sec. 5-5c]

Guggenheimer, K.: *J. Phys. Radium*, **5**:253, 475 (1935). [Sec. 2-5]

Gurney, R. W., and E. U. Condon: *Nature*, **122**:439 (1928); *Phys. Rev.*, **33**:127 (1929). [Sec. 4-5b]

Hahn, O., and F. Stassmann: *Naturwiss.*, **27**:11, 89 (1939). [Secs. 5-1, 5-7]

Haxel, O., J. H. D. Jensen, and H. E. Suess: *Phys. Rev.*, **75**:1766 (1949); *Z. Physik*, **128**:295 (1950). [Sec. 2-5c]

Heisenberg, W.: *Z. Physik*, **43**:172 (1927). [Sec. 2-2f]

———: *Z. Physik*, **77**:1 (1932); **78**:156 (1933). [Secs. 1-1, 1-2a, 6-2]

Heitler, W.: "The Quantum Theory of Radiation," 3d ed., Clarendon Press, Oxford, 1954. [Sec. 3-4a]

Hofstadter, R.: *Phys. Rev.*, **75**:796 (1949); see also J. H. Neiler and R. E. Bell:

The Scintillation Method, in K. Siegbahn (ed.), "Alpha-, Beta-, and Gamma-Ray Spectroscopy," chap. 5, North-Holland Publishing Company, Amsterdam, 1965. [Sec. 3-6]

——— (ed.): "Electron Scattering and Nuclear and Nucleon Structure," W. A. Benjamin, Inc., New York, 1963. [Secs. 1-1, 1-2b, 6-2]

———, F. Bumiller, and M. Croissiaux: *Phys. Rev. Letters*, **5**:236 (1960). See also R. Hofstadter, 1963. [Sec. 6-2]

———, H. R. Fechter, and J. A. McIntyre: *Phys. Rev.*, **92**:978 (1953). See also R. Hofstadter, 1963. [Sec. 1-1]

Kaplan, I.: "Nuclear Physics," 2d ed., Addison-Wesley Publishing Company, Inc., Reading, Mass., 1962. [Sec. 3-3]

Kittel, C., W. D. Knight, and M. A. Ruderman: "Mechanics-Berkeley Physics Course," vol. 1, McGraw-Hill Book Company, New York, 1965. [Sec. 2-2a]

Lee, T. D., and C. N. Yang: *Phys. Rev.*, **104**:254 (1956); **105**:167 (1957). [Secs. 1-1, 4-6h]

Littauer, R. M., H. F. Schopper, and R. R. Wilson: *Phys. Rev. Letters*, **7**:144 (1961). [Sec. 6-2]

Marion, J. B., and L. L. Fowler (eds.): "Fast Neutron Physics," Interscience Publishers, New York, 1963. [Sec. 5-7c]

Mayer, M. G.: *Phys. Rev.*, **75**:1969 (1949). [Sec. 2-5c]

Meitner, L., and O. R. Frisch: *Nature*, **143**:239 (1939). [Sec. 5-1]

Mott, N. F.: *Proc. Roy. Soc.* (*London*), **A126**:259 (1930). [Sec. A-5]

Pauli, W.: "Rapports du 7^e Conseil de Physique Solvay, Brussels, 1933," Gauthier-Villars, Paris, 1934. [Sec 4-6a]

Planck, M.: *Ann. Physik*, **4**:553 (1901). [Sec. 2-2a]

Powell, C. F. (1946): see C. M. G. Lattes, H. Muirhead, G. P. S. Occhialini, and C. F. Powell: *Nature*, **159**:694 (1947). [Sec. 1-1]

———, P. H. Fowler, and D. H. Perkins: "The Study of Elementary Particles by the Photographic Method," Pergamon Press, London, 1959. [Sec. 3-2]

Reines, F., and C. L. Cowan, Jr.: *Phys. Rev.*, **92**:830 (1953); *Nature*, **178**:446 (1956); *Phys. Rev.*, **113**:273 (1959); see also F. Reines, Inverse Beta Decay, in K. Siegbahn (ed.), "Alpha-, Beta-, and Gamma-Ray Spectroscopy," chap. 24, sec. H, North-Holland Publishing Company, Amsterdam, 1965. [Sec. 4-6g]

Robinson, B. L., and R. W. Fink: *Rev. Mod. Phys.*, **32**:117 (1960). [Sec. 4-6f]

Rutherford, E.: *Phil. Mag.*, **21**:669 (1911). [Sec. 1-1]

———: *Phil. Mag.*, **37**:581 (1919). [Secs. 1-1, 5-1]

———, J. Chadwick, and C. D. Ellis: "Radiations from Radioactive Substances," Cambridge University Press, London, 1930. [Sec. 4-1]

——— and F. Soddy: *Phil. Mag.*, **4**:370, 569 (1902); **5**:576 (1903). [Secs. 1-1, 4-5, 4-6]

Schiff, L. I.: "Quantum Mechanics," 2d ed., McGraw-Hill Book Company, New York, 1955. [Secs. 2-2c, 2-2g, 2-2h, 2-5b, 4-4c, 4-6c, A-2]

Schrödinger, E.: *Ann. Physik*, **79**:361, 489, 734 (1926); **80**:437 (1926); **81**:109 (1926). [Sec. 2-2a]

Schwinger, J., and E. Teller: *Phys. Rev.*, **52**:286 (1937). [Sec. A-4]

Segrè, E.: "Nuclei and Particles," W. A. Benjamin, Inc., New York, 1964. [Secs. 3-3, 3-3a, 3-3b, 3-5b, 4-6c]

Siegbahn, K. (ed.): "Alpha-, Beta-, and Gamma-Ray Spectroscopy," North-Holland Publishing Company, Amsterdam, 1965. [Sec. 3-6]

Smith, C. M. H.: "A Textbook of Nuclear Physics," The Macmillan Company, New York, 1965. [Sec. 2-3d]

Sternheimer, R. M., in L. Marton (ed.): "Methods of Experimental Physics," vol. 5, part A, Academic Press Inc., New York, 1961. [Sec. 3-2]

Thomson, J. J.: "The Corpuscular Theory of Matter," Constable and Company, Ltd., London, 1907. [Secs. 1-1, 1-2b]

———: *Phil. Mag.*, **24**:209 (1912); *Proc. Roy. Soc.* (London), **A89**:1 (1913). [Sec. 1-2a]

Weisskopf, V. F.: *Science*, **113**:101 (1951). [Sec. 2-5]

———: *Rev. Mod. Phys.*, **29**:174 (1957). [Sec. 5-1]

von Weizsäcker, C. F.: *Z. Physik*, **96**:431 (1935). [Sec. 2-4]

Wheeler, J. A.: Channel Analysis of Fission, in J. B. Marion and J. L. Fowler (eds.), "Fast Neutron Physics," vol. 2, p. 2051, Interscience Publishers, New York, 1963. [Sec. 5-7e]

Wigner, E.: *Phys. Rev.*, **43**:252 (1933); *Z. Physik*, **83**:253 (1933). [App. A]

Yukawa, H.: *Proc. Phys. -Mat. Soc. Japan*, **17**:48 (1935); *Rev. Mod. Phys.*, **21**:474 (1949). [Secs. 1-1, 6-2]

269